The Energies of the Day

The Predictions of the Sacred Maya Calendar for
2012

Denise Barrios
Denise De Peña

Mystic Maya Publications™

THE ENERGIES OF THE DAY 2012 - The Predictions of the Sacred Maya Calendar.
Copyright 2012 © Denise De Peña and Denise Barrios.

Mystic Maya Publications ™
San Pedro El Alto
La Antigua, Guatemala
www.mystic-maya.com

Cover and text design by Denise De Peña
Photographs: Denise De Peña
Edited by Denise De Peña and Denise Barrios

ALL RIGHTS RESERVED. No part of this book may be reproduced in any manner whatsoever without the written consent of the authors.

Printed in the United States of America.

This Book Belongs To

On the Cover

Pass your hands over the maize
                    over the tz'ite'
so that it may be done
so that it may be clarified if we should chisel, and
                if we should sculpt its mouth,
                          its face in wood, the priests were told.
Therefrom the count of days
            of creation, was cast
                    was divined on the maize
                            on the tz'ite'.

—Popol Wuj

This passage in the sacred book of the Maya, the Popol Wuj, relates how the *tz'ite'* and maize seeds were used in a divination ceremony for the creation of humankind. The mythical accounts of the Popol Wuj are at least 2600 years old. To this day Maya daykeepers pass their hands over the *tz'ite'* when conducting this ancient divination ritual. The photograph on the cover of the book depicts a handful of the magical *tz'ite'* beans, and the colors on the cover relate to the symbolic colors of the four cardinal points as described in Maya mythology: red for the east, black for the west, white for the north, and yellow for the south. Each has a specific energy and meaning.

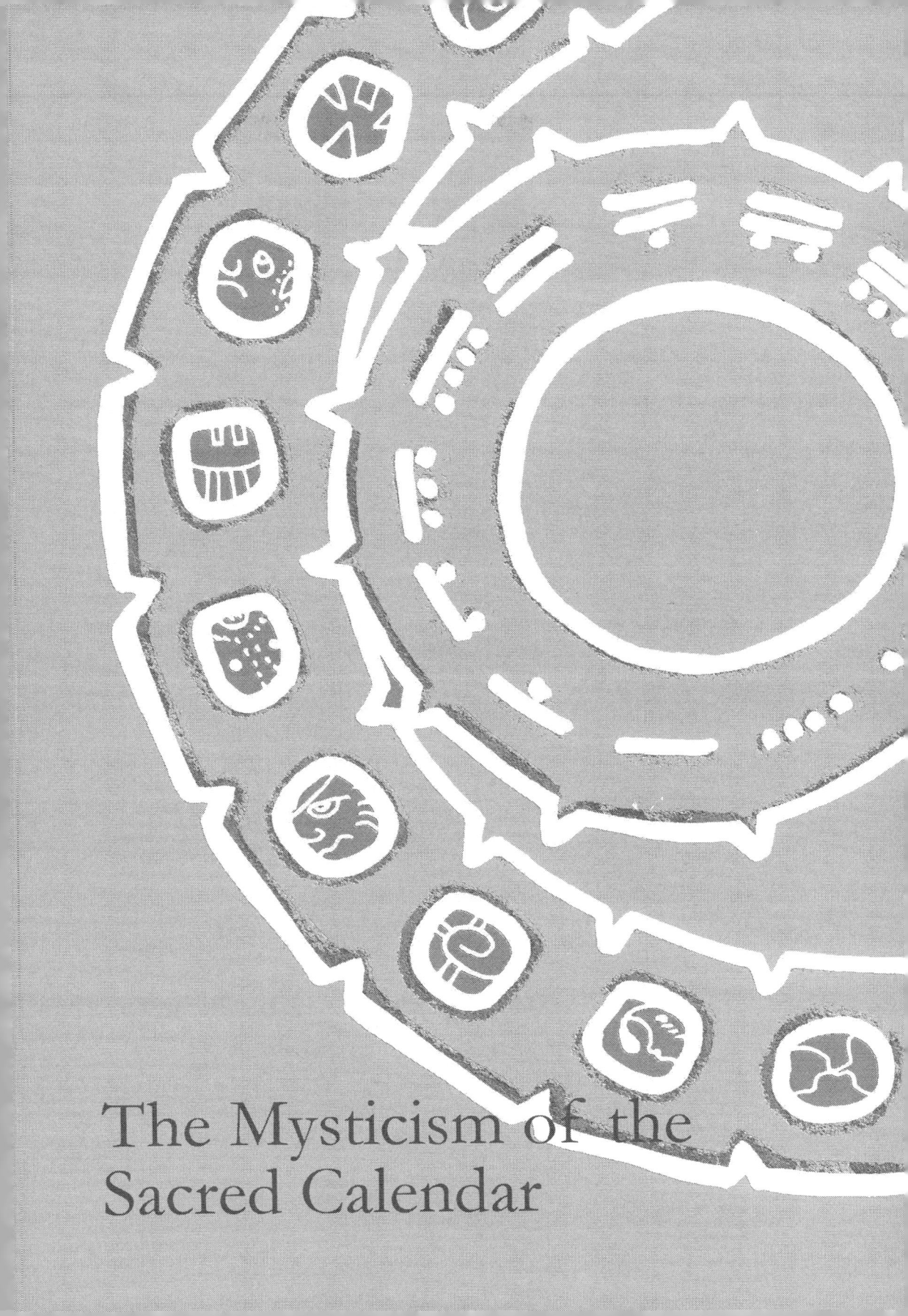

The Mysticism of the
Sacred Calendar

The Cholq'ij

The *Cholq'ij*, which means "the count of days," possesses magical and divinatory qualities that the *Ajq'ij'* —the spiritual guide and "day keeper"— uses to predict the future and interpret life's events. The *Ajq'ij'*, as the guardian of tradition, records the passage of time, since it contains the story of his people and, according to the vision of the Maya Grandmothers and Grandfathers, all that has ever occurred will occur again, as the spiral of time and space repeats in cycles without end.

Also known as the *Tzolkin*, or Sacred Calendar, it is the guide and axis around which all the activities of the Maya people revolve, as their mythology, rites, ceremonies, social events, and agricultural activities are completely bound to this calendar. More than just a calendar that marks the passage of time, The *Cholq'ij* also renders an account of the energies that influence daily life.

The 260 days of the *Cholq'ij* are divided into 13 periods of 20 days. The number 20 arises from the convergence of the 10 cosmic and 10 telluric energies of the fingers on the hands and feet. Each of the 20 days has a distinct energy and is represented by a glyph, or *Nawal*. Likewise, each *Nawal* relates to one of the three levels of the Maya Universe; that is, the underworld, the earth, or the heavens, and to one of the four elements: fire, earth, water, or air.

Additionally, the 20 *Nawals* combine with other energies known as Powers, which run from 1 to 13. The 13 Powers are linked to the energetic load of the 13 main joints of the human body. In this way, the energies of each day are formed by the confluence of the energy of a Power and the energy of a *Nawal*, and this is what determines the strength and the character of the energy that each of the 260 days of the *Cholq'ij* will display.

The *Nawals* and the Powers run together continuously and parallel in such a way that the combination of a Power with a *Nawal* will not repeat for 260 days. A calendar cycle is complete when each of the 13 Powers has combined with each of the 20 *Nawals*. In this way, the *Cholq'ij* has recounted the days for thousands of years and has begun each new cycle of 260 days on the day 8 B'atz', which is when the Maya New Year is celebrated.

Each day possesses a unique energy that essentially serves to strengthen our spirituality, provides us with ethical and moral guidelines, and leads us to better know ourselves and be better persons. Its symbols, colors, and numbers express feelings, values, and knowledge that gives meaning to our spiritual and material lives, because it expresses life itself in all of its dimensions.

The *Cholq'ij* is also tool that allows us —through the Maya Cross that is thereon configured for each of the 20 *Nawals*— to determine the energy that will govern our lives from the moment of conception and birth. It helps us gain better understanding of our own nature, expand and take advantage of our abilities, correct our negative aspects to attain a balanced and harmonic existence, and reach our full potential.

Its 260 days concur with the time it takes the moon to orbit nine times around the earth and the time a child remains in her mother's womb. Consequently, the *Cholq'ij* is a lunar calendar and its energy is feminine, as nine is the number of feminine energy and the number at the peak of the woman's pyramid.

The Sacred *Cholq'ij* Calendar is part of the longer *Ab'* calendar. Also known as the "long count" calendar, the *Ab'* is a solar calendar whose energy is masculine. Its 360 days are divided into 18 periods of 20 days, plus a 5 day period known as the *Wayeb*. Both calendars coincide every 52 years, from where the calendrical rotation begins anew at the initial position.

2012: The Door to Transition
What it Means and How you can Prepare for this Prophetic Year

Maya oral tradition tells us that we are the fourth civilization that has reached this moment and tries to transcend to the next level: the Fifth Sun. The first civilization had feminine energy and was governed by the element Fire. The second civilization had masculine energy and was governed by the element Earth. The third civilization again had feminine energy, and was governed by the element Air. The fourth, to which we belong, has masculine energy and is governed by the element Water. At the present time we will transcend to a new epoch in which there will be balance between both masculine and feminine energies, and this era will be governed by the fifth element: the Ethereal Element, which embodies spiritual perfection.

So then, what is the significance of December 21, 2012? It is the end of a cycle in which a subtle change of energy will take place and allow humankind to enter into a more spiritual vibrational state; one of greater consciousness in which we will reevaluate our priorities and recapture our intimate connection with nature. Some may not even notice this transformation; they may think that December 21 is a day like any other, but a change in energy at the spiritual and sensitive level will take place, and those that are prepared will have no difficulty in recognizing it.

Mother Earth —like the living being that she is— will elevate her spiritual condition. This is why humankind, as a manifestation of her evolutionary legacy, will be able to transcend to the next level. Sadly, we have continued to destroy our Mother Earth and have put our own evolutionary history at risk, just as the previous civilizations did when they reached this point. Notwithstanding, once again we have the opportunity to ascend to a higher level, and to perfect our spiritual evolution through Her.

Wayeb, a Period of Introspection

The first thing you can do to prepare for this moment is to understand who you are. Western culture does not teach us how to find ourselves or to develop the capacity to recognize who we really are and where we are headed. The first steps are to make up your mind to look deeper within yourself, to seek harmony with nature, and to analyze before acting in order to conduct your life wisely.

In Maya culture people examine their life every year during the last month of the solar *Ab'* calendar, the five-day month known as the *Wayeb*. The *Wayeb* is a period of transition that is known in the West as the unlucky or lost days, but in the traditional Maya worldview it is a time for introspection and gratitude to the *Ajaw*. It is the space and time in which we can be with ourselves, review our actions and the positive or negative impact they may have had on our life, as well as the purpose of our existence. This is a practice that you can adopt, and it is not necessary to wait for the *Wayeb*. However, if you want it to be effective you must make it your own, as the purpose of this exercise is to find your essence.

One of the techniques consists of recapitulating and focusing on the most important points, which are primarily your relationships with others:

Your romantic relationships. Start by reevaluating your most recent relationship and going back in time to the first one. Our love history usually has the greatest impact on our lives. Accept it and embrace it just as it is as impartially as possible. In this way, you will be able to understand where you have succeeded or failed, and your responsibilities to the people that have been your partners. Intend from now on to be a loyal, honest, and loving person, just as you would like your partner to be with you.

Family. Family is very important, as it is the context in which you grew up, and from which the majority of your traditions, attitudes, and prejudices might spring. Examine and liberate yourself of the prejudices and pointless attitudes. Frequently, they arise from the mistaken notions we have about others and, therefore, they lack any real validity, but they halt our spiritual development. Think about what you can do to develop stronger bonds with the members of your family and reaffirm your traditions, as they lend a sense of belonging and continuity to your life.

Your general environment. This includes your immediate surroundings: your workplace, friends, your life at school… any event that may have marked your life, whether it happened with your intervention or without it. Examine your role in the life of others, and the ways in which you can become a generous person that is sensitive to their feelings. This circle is, generally speaking, where we discover our life's purpose. Open up to its energies.

- REVIEW each of these circles until your memory becomes clear to the point in which you can remember all of the details of the events that have had an impact on your life.
- DETERMINE whether the success you've earned throughout your life fulfills you; is it really what you expected to achieve?
- ACCEPT the things that have happened in your life; accept the mistakes and understand that everything is part of your evolution. Accept yourself just as you are!
- DEFINE what it is that you need to do to redirect your life: What makes you happy or unhappy? What do you want from life? What is keeping you from achieving your goals? And, what does "real" success mean to you? Then,
- DECIDE where it is you want to head; determine the strengths you have that will help you reach your goal, and the weaknesses that are stopping you now. Which is the road that you must take to reach the spiritual, emotional, social, work, financial, or family-related goals that you desire? And, finally,
- PUT IT INTO PRACTICE. Intentions without actions will have no positive outcomes in your life.

Undertake your analysis based on your true potential, so as not to become frustrated by trying to achieve something that may be impossible. In 2012 the *Wayeb* begins on February 17 and ends on February 21.

Meditating through Fire

Another way to prepare for this date is by meditating in front of a candle. Fire wields an impressive power over us, as we evolved throughout history in its company. Fire allowed us, as we know, to survive the winters and cook our food. It also became the center of our homes. Fire was the sacred presence around which all the members of the household gathered until artificial fire, electricity, came into our lives.

In former times the family would gather around the fire and communicate with each other, which strengthened family bonds and fostered unity. Today, televisions, computers, smartphones and other technological devices have come between families and communities. We have become a self-centered society.

Meditating in front of a candle and concentrating on its flame will allow you to connect with yourself, the cosmos, and everything around you, and fill you with the energy of Mother Earth's inner fire —*Kak Alom*— or the magma that is her heart. The candle's flame symbolizes your spirit as it connects to all of creation and to the *Ajaw*, the Creator. While meditating and focusing on the fire, visualizing its flame in your pineal gland, your mind will lose its inflexibility. In this way, it will allow you to tear down any rigid structures and open the necessary space for clarity to enter, help you connect to your inner self, purify you, and assist the development of your spiritual powers.

As you become aware of the significance of life itself you will find your mission, and acquire the power and strength to help Mother Earth transcend to the new era. Additionally, you will serve as an inspiration to others and attract more people toward a spiritual, communal, and ecological lifestyle, and thus contribute toward the balance of life on this earth. Take into account that spirituality today demands action.

You can also use a bonfire or the energy of the sun at dusk in place of the candle and flame to meditate.

Meditating on the Nawals

Each of the 20 *Nawals* contains and represents an energy of its own and a special power. Focusing on each of them can help awaken your dormant powers and activate your inner knowledge. These are the energies they transmit:

B'ATZ'	The power to heal and self-expression through art.
E	The power to connect with the energies of the underworld, the world, and the cosmos through *Saq'be*, the Sacred Road.
AJ	Clairvoyance and telepathy. Aj grants spiritual authority and the capacity to receive and transmit sacred messages.
I'X	Cunning, agility and high magic.
TZ'IKIN	Vision, intuition, and revelations in dreams.
AJMAQ	Awakens consciousness and enlightens the mind.
NO'J	Transforms knowledge and experience into wisdom. Grants clarity.
TIJAX	Cuts out negativity. Allows opening new paths and connecting with the energy of lightning to ask for revelations.
KAWOQ	Knowledge on medicinal plants. The ability to work with others in a group.
AJPU	The strength of the Spiritual Warrior that overcomes all obstacles.
IMOX	Projects the strength and purity of water. Strengthens internal powers and confers intuition.
IQ'	Nourishes and renews the mind, the body, and the spirit. Develops the powers of the mind and perception.
AQ'AB'AL	Reveals the mysteries of life. Brings light into darkness.
K'AT	The ability to unravel problems.

KAN	Awakens the inner serpent that lies at the base of the spine and creates a link with the cosmic and telluric energies.
KAME	Communication with superior beings and ancestors, so they may guide us and awaken our ancestral knowledge.
KEJ	Presides over the four elements. Grants strength and balance, and connects us with Mother Earth's wisdom.
Q'ANIL	Allows our inner wisdom to flourish.
TOJ	Understanding the signs of the *Tojil*, the Sacred Fire.
TZ'I'	Grants justice and allows us to see the truth. The ability to communicate through writing.

Light a candle for each of the four cardinal points: red in the east, black, purple or deep blue in the west, white in the north, and yellow in the south. Sit in the center and light a candle that represents the *Nawal* of the day, on which you will focus. The best time of the day to do this is at dusk, as the energy of the sun and day converge with the energy of the nighttime heavenly bodies. Quiet your mind and visualize the sign of the *Nawal*; feel how it flows within you until it alights between your eyebrows. Meditate on its meaning, allowing it to transmit its teachings and provide you with the powers you need at this moment in time.

Meditation during Solstices and Equinoxes

Solstices and equinoxes are very important to the Maya culture, as they change the prevailing energies at planetary level. Maya Grandfathers and Grandmothers liken them to a hammock, which upon reaching its highest point at one extreme swings back to the center and to the other side, and this is precisely what happens during solstices and equinoxes. When we are at these extremes we are more open to perceiving the energy, as we are more sensitive to it. This reaction increases if we are at a sacred location in which there is an energetic convergence, such as at a Maya pyramid or at the sacred sites of other tribes. If you are at one of these sites during a solstice

> **Take advantage of the activating energy of a solstice**
>
> **... and the balancing energy of an equinox**

or equinox, you will find that you can establish a connection with the cosmos that will show you the portal through which the terrene can elevate to the supreme.

During a solstice or equinox (preferably while you are on a Maya pyramid or at a sacred site) you will witness *Kukulkán* —the plumed serpent— descending from the cosmos and merging with the energy of the earth and the underworld. At this moment, focus your meditation on *Kukulkán* positioning itself on your spine, activating your inner energy, and creating a beam of light that will bond you with the world and the cosmos. This experience will help awaken your consciousness and unveil the true meaning of your existence, so that in harmony with everything that surrounds you, you can realize the divine plan that was created just for you, and thus attain the truest joy.

The Energies of the Maya Calendar Glyphs

The Nawals

Each of the 20 days of the *Cholq'ij* calendar is governed by a *Nawal*, which is an energy or spirit of nature, and its corresponding element: water, fire, earth or air. In this way, for example, the *Nawal* for I'x is the jaguar and its element is water, the Nawal for Aq'ab'al is the bat and its element is fire, and No'j —which is governed by the coyote— has the element of air.

The interpretation of the energies of the day glyphs is based on the intuition, knowledge and sensitivity of the *Ajq'ij*, who is the person that deciphers them. Although they repeat on a cyclical basis, it is never a mechanic interpretation.

Following are the 20 *Nawals* and their glyphs, and an interpretation of their energies.

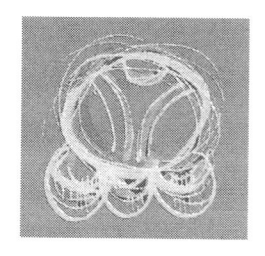

B'atz'

B'atz' is the thread of destiny. Everything that has occurred to us in the past, anything that is happening to us now, and everything that will occur in our future —and whatever has or will occur in the history of the world— is the essence of B'atz'. B'atz' exists so that creation may fulfill its fate. It is the thread of time that unfolds before us and weaves our histories; it is the beginning, the start of life, and original wisdom. The upper part of the glyph depicts a cone with time rolled up within it. As it unrolls downward it passes underneath the earth toward the outer angles, which represent male and female antithesis.

E

The path that leads us to our precise destination in life is E. Time begins with B'atz' and ends with T'zi', but E is the energy and the means by which we will travel our road. It is the sacred road of life; it sets the conditions of our journey and provides us the power to accomplish the mission for which we were born. As well, E is the stairway that spans between the earth, the heavens, and the underworld, and guides us in our travel between these points. The right upper corner of the glyph depicts an ear that represents the path, and the dots are the stones in our path. On the left, the teeth represent the stairway of our spiritual journey.

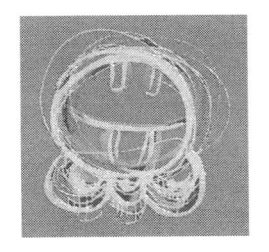

Aj

The energy of Aj embraces spiritual law, heavenly authority, and the Divine Justice that radiates from the sacred altars and the staff of worldly authority. It represents the pillars of inner strength and the columns that sustain the family home. Thus, its energy overflows onto the children and everything that relates to the home. On the day of Aj the sacred maize and animals were domesticated. It personifies integrity, honesty and abundance, and is symbolically embodied in the sugarcane and the cornstalk. On the glyph, the small vertical lines at the top symbolize spiritual growth and plenteousness. The horizontal line represents the horizon, the altar table, and the sacred trees whose canopies are altars.

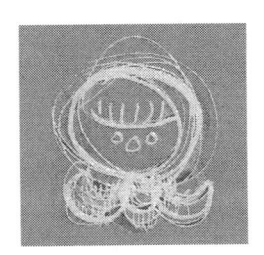

I'x

I'x possesses the power to expand our mind, and it gives us the strength with which to reach high levels of consciousness. I'x is the mother, the wife, and Mother Earth; thus, I'x is mystical feminine energy, and from her springs forth the creative impulse of the Universe. I'x is also the feline energy that possesses the jaguar's strength, power, and cunning. The mountains, hills, and plains, and all the other sacred places, are under the protection of the *Nawal* I'x. On a day of such high magic the impossible is possible, and the unreal may be real. This glyph represents the vital core of the planet, the feminine reproductive organs, and the face of the jaguar. The dots are a map of the energies of the world.

Tz'ikin

Tz'ikin, the eagle, personifies liberty and the sacred vision. It symbolizes the panoramic vision of the eagle, one that is free from the constraints of time or space, and it has the power to mediate between the Great Father, *Ajaw*, and mortals. The energy of this sign awakens global consciousness and instills idealism and community service. It also possesses a magnetic energy that attracts material wealth, love, luck, and success; that is, the fulfillment of all human desires. The top of the glyph depicts the pate of the bald eagle, and the vertical lines represent the feathers at the back of its head.

Ajmaq

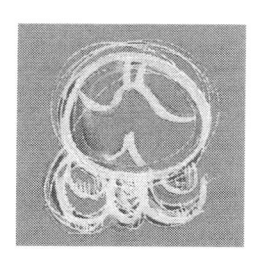

Ajmaq is the wisdom of the Elders, fruit of the thirst for knowledge and experience. In this way, Ajmaq transmits the virtue of a long and honorable existence but is a day for repentance, forgiveness, and reconciliation with the Great Father. It is the energy that governs harmony and discord, and allows us to recognize our innermost essence and degree of spiritual evolution. Through the energy of Ajmaq we can grow to become role models and thus be greatly blessed. The lines in the glyph that extend in every direction stand for the spiritual communion between the mind and the *Nawals*, a mind that is in a state of forgiveness and enlightenment.

No'j

No'j is the energy that transforms our knowledge and experience into wisdom and spiritual understanding through its powerful connection with the Universal Mind. It is the interaction between the immeasurable and the everlasting; it is freedom from the rigid boundaries of reality as conceived by our human minds. The Maya Elders meet in council under the invoked protection of No'j because we humans have knowledge but lack wisdom, which can only be granted by *Ajaw* through the energy of No'j. The space-time mystery is within the province of No'j, which grants enlightened spirits the ability to travel with the mind. At the center of the glyph are three dots that represent the three degrees of ever-increasing spiritual achievement; the line in the middle depicts the human intellect.

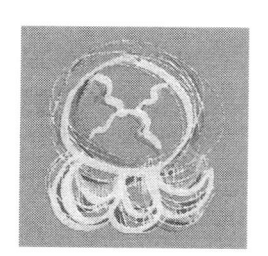

Tijax

The double-edged obsidian knife, Tijax, embodies the trials that we must overcome and the energy that heals. It is the *Nawal* of healers, and it has the power to cut out all ailments and tribulations, as it possesses the power of the vitrified heart of Mother Earth —the obsidian that was created by the Sacred Fire. And so, the energy of Tijax helps us cut through the mysteries of life and the advent of positive and negative forces. Its strength springs from thunder and lightning, purifying celestial forces that, upon converging, open before us new roads and dimensions. The glyph symbolizes the borders of the obsidian knife as seen from its tip.

Kawoq

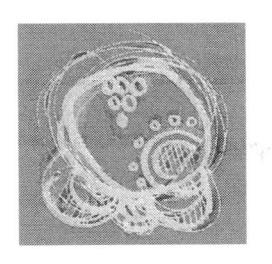

The energy of Kawoq presides over matters that concern the family, the community, and society overall, and it bids that our deeds should benefit our communities. It is the passage of time —the cluster of days— and the lessons that each of these days bring throughout the different phases of our life. As well, it projects its energy onto the planet, the solar system, the galaxy, and the universe and its expansion, and it symbolizes the strength in union, global consciousness, and the evolution of the Cosmic Plan. The spheres at the top of the glyph represent the members of the family, the families in the community, and the communities of the world. The crosshatched lines symbolize cooperation toward accomplishing shared goals.

Ajpu

This is the day of the Great Father, *Ajaw*, whose grandeur and vitality is projected in his solar likeness. Ajpu is the energy of the blowgun hunter, the Spiritual Warrior represented by the mythical twins *Jun Ajpu* and *Ixb'alamke*, who died and were reborn after surmounting many spiritual tests —an allegory for spiritual awakening and the triumph of good over evil. It also references solar and lunar eclipses, and the mystical ball game that sustains the cosmic order of the universe and the regeneration of life. Ajpu, then, is the hunter, warrior, and unwearied traveler; the force that can help us overcome negative energies and triumph through vision and spirituality. The glyph depicts the face of the hunter with a blowgun in his mouth.

Imox

Imox is the energy that governs change. It represents the innermost recesses of the mind and the interconnection of ideas. Because of its influence on our unconscious mind it relates to everything that is subtle, insightful, and eccentric. It is like the energy of scattered water that does not allow us to reach our goals, but if we can channel it, it gives us the power to create whatever we set our minds to achieve. It is an energy that can lead to confusion and insanity, and give rise to strange events. It presides over rain, rivers, lakes and oceans, and the creatures that dwell there. Thus, it epitomizes the primordial source of life and grants us the ability to understand the messages of nature. This glyph is the image of rain, and a gourd that is filled from the top with water; the vertical lines represent its fullness. As well, it depicts the full breast of a milking mother.

Iq'

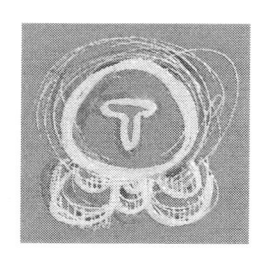

The element that governs our ideas is the wind, as embodied in the energy wielded by Iq'. It represents the divine breath of life; the wind of change that activates the energy within our body and inspires us into action. It has the virtue of purifying our body, mind, and spirit, and nourishing our thoughts. On account of its impalpable nature, Iq''s energy is pure and crystal clear, and has the power to harbor visions and bring forth beauty and harmony. Iq' is also the invisible space between matter and the expanse between heaven and earth. At its center the glyph depicts a window like those found in Classic Era Maya architecture. These windows often had giant flutes that made music in the wind.

Aq'ab'al

Sunrise and sunset, light and shadow, the two sides of a coin; such is the energy of Aq'ab'al: opposite yet harmonious. Who can say when night ends and day begins? The renewing energy of this sign helps us unfasten our moors and head toward new destinations as we experience individual, chronological, and cyclical changes during our lifetime. Aq'ab'al signals a new dawn and is the force that helps us end the monotony that sometimes pervades our existence. In this way, Aq'ab'al can be interpreted as the youthful guiding energy of our daily lives that gives us hope and offers us solutions. The line that splits toward the bottom of the glyph represents the light that shines at dawn and dispels materialism from the face of the earth.

K'at

K'at is the energy that tangles us in a web, mostly of our own making, or in the traps that life sets out to instruct us. It can be physical, mental, emotional, or spiritual bondage and oppression, but it also signifies experience and the life changing events that can liberate us. Just as it can ensnare us in its negative energies, it can also group and bring together the people and the elements that can liberate us from the obstacles ahead if we learn to cast its net at the right time. As well, it is the delicate netting that can hold in our memory everything that we learn. The glyph represents the earth that is trapped by gravity, as depicted by the notch at the bottom that divides the base into opposing forces.

Kan

It is the mission of every human being to ignite Kan, the serpent of fire whose energy resides at the base of the spinal column. It is the force of life that is manifested as sexual energy; it is the helix that contains our genetic code and humankind's collective memory —it is every form of evolution. Kan is the energy of justice, truth, and peace; it stands for the cycles of time, the transmutation of knowledge into wisdom, and for sexual magic. It is balance, power, and authority. This day has a very strong energy because, additionally, it carries the energy of the anger that the mythological *Ajpop Katuja* took with him when he descended to the underworld. The stylized glyph represents the crosshatched skin of the plumed serpent *Kukulkán*.

Kame

Only at death can we return to our point of origin and find peace and harmony. Kame is the energy of our spiritual and family lineage; it is the face of death and communication with those who dwell in the other dimension. It has the power of our ancestors, who invisibly accompany and protect us throughout life; woe to them that do not keep their memory, they will have no guidance! Kame is the energy of the revelations that are induced by the seven sins: ignorance, pride, ambition, envy, ungratefulness, untruth, and wrongdoing. It is obscurity, fear, shame, and suffering. This energy dissipates all things —good and bad— and can weaken or strengthen a person's powers. The closed eye and protruding teeth depicted on this glyph clearly portray death.

Kej

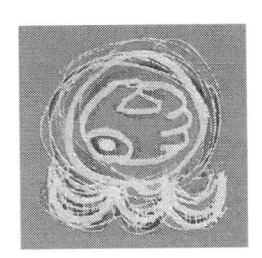

Kej projects the energy of four, and as such, balance. It is represented in the four legs of the deer that stretch to the four cardinal points and sustain the four pillars of heaven and earth; it is the four paths, the four elements —earth, fire, wind, and water— and the four manifestations of the human being: physical, mental, emotional, and spiritual. Kej is the energy that protects the natural world and is called to maintain balance between humans and Mother Earth. It embodies profound teachings and advice, and the forces that sustain human existence. The touching fingertips depicted in the glyph denote the closed circuit of energy that charges the four cardinal points.

Q'anil

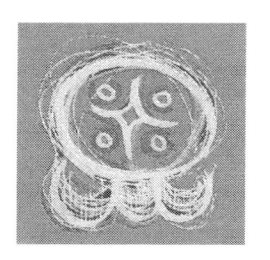

Q'anil is the energy of creation; it is the seed of life in all of its manifestations and eternal regeneration. This is the day *B'itol* and *Tz'aqol* sowed the seeds of life in our corner of the universe. It is a day of abundance and shared harmony, which are the fruits of love and understanding. Q'anil endows us as well with the spiritual seeds that contain the codes we need to conduct our lives. At its center the glyph depicts the hole in the earth that is left by the farmer's planting stick, and the four small circles represent the heirloom white, yellow, black, and red corn kernels, and the four cardinal points.

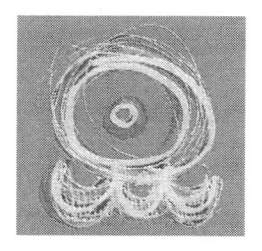

Toj

The energy of Toj is expressed through cause and effect, action and reaction. It is a day in which we must repay Mother Earth, and the Creator and the Maker —*B'itol* and *Tz'aqol*— for what we may have done, for what we are, and what we have received, with a Sacred Ceremonial Fire, which is the manifestation of *Ajaw* upon the earth. Toj then, is the energy that conducts light, and through the Sacred Fire we can petition for all our needs and reconcile with the Father. This fire has the power to cure all our ills and free us from any negative energy that we may have. The glyph symbolizes the strength of Father Sun; it is a wheel and axis that represent cause and effect.

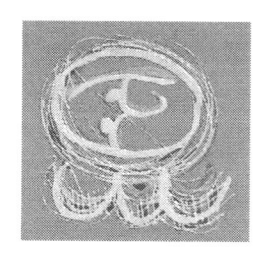

Tz'i'

Tz'i' embodies the energy of divine and natural law under which we all live; it is the righteous path and the sign of authority and justice. The dog is an assistant and counselor that represents the Creator on earth, and it is he that is charged with applying justice to insure that it is served. Thus, truth always comes to light on this day. It imbues special people and places with the energy of the cosmic authorities, and it is bestowed on those that govern us. The energy of Tz'i' also relates to the written word, especially to sacred writings. This glyph symbolizes the staff of authority that is used by indigenous officials and the tail of the dog.

The Energies of the 13 Powers

The distinctive way in which the Maya wrote their numbers, and the fact that the energies of the Powers reach only up to the 13th level, correlates directly with our body and the points of entry through which energy can enter.

On the one hand, the 13 Powers are linked to the energetic charges of the 13 main joints in the human body, thus: two ankles, two knees, two hip joints, two wrists, two elbows, two shoulder joints, and one neck joint.

On the other, the graphic representation of the numbers is based on our own hand. In this way, the dot, which represents the unit (1), is the point of our finger as it would leave a mark on a piece of paper, and the bar, which represents five units (5), is the representation of our vertically extended fingers as seen from the side. Maya numbers can be written either vertically or horizontally. For example, the following figures represent the number 13 (vertically and horizontally), number 1, number 5, and the day 13 Imox.

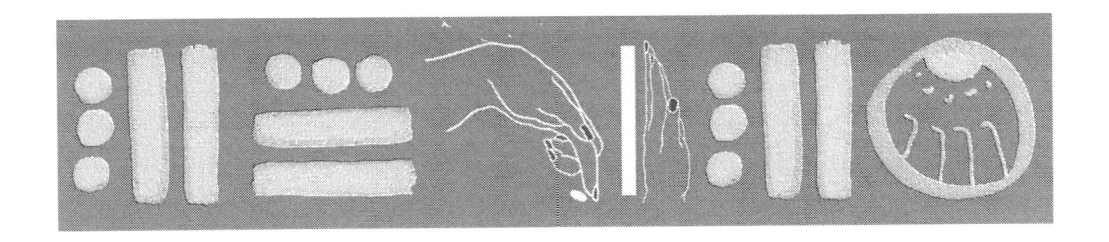

To interpret the energetic charge of a specific day, the energy of each Power is combined with the energy of each corresponding day sign (*Nawal*, or spirit of the day), and thus, it gives us a special reading for each day. As there are 20 *Nawals*, the combination of the energy of each Power with the energy of each *Nawal* will repeat only once every 261 days.

The Names and Energies of the 13 Powers

Power 1 ~ *Jun* represents unity and the beginning. This energy contains force and wisdom, but lacks experience. It is a day in which creative and innovative solutions may arise, and an auspicious day to begin new projects or make significant life changes. It stands for the essence of all things.

Power 2 ~ *Ka'i'* alludes to duality and union. It is a favorable day to establish partnerships and sign contracts. A life partner or other special person may come into your life on this day. It is also a good day to improve relationships with family and friends. It stands for the choices you must take.

Power 3 ~ *Oxi'* symbolizes communication, creativity, and optimism. It is represented by the three cosmic stones that were placed at the center of the Cosmos at the beginning of Maya creation. Use your imagination and talents to the fullest on this day. It stands for the results of your efforts.

Power 4 ~ *Kaji'* projects stability, firmness, and security. It possesses the sturdiness of the deer that stands on its four legs, and the force and wisdom of the four *Balam'eb* (men-gods). It alludes to the four cardinal points, the four races of humanity, the four colors of corn, and the four elements. Personal or business projects started today will be sound and prosperous. It stands for structure and order.

Power 5 ~ *Woo* signifies liberty, charisma, magnetism, and luck. It is a good day to make changes. Because of its lucky nature, on this day you are more likely to find love, take unexpected trips, or close a good deal. Today you can achieve anything you set your mind to. It stands for the love that comes into your life and flows from you.

Power 6 ~ *Waai'* entails learning, changes, and responsibilities. It is a good day to undertake small changes in your life; a day in which to thank the *Ajaw* (Supreme Being) for all that you have received. Strive to give back to your community and resolve any pending issues. It stands for life's trials and personal development.

Power 7 ~ *Wuqu'* projects faith, spirituality, intuition, analysis, and introspection. This is a good day for spiritual development, for self-reflection, and for internal growth. It is not the best day to make important decisions. It stands for balance and possesses the power to impel.

Power 8 ~ *Waqxaqi'* stands for authority, ambition, and materialism. This day is filled with physical energy and is auspicious for men. It brings excellent opportunities in the financial area and is a good day for curtailing expenses and increasing savings.

Power 9 ~ *B'eleje'* reflects feminine energy, love, wisdom, and spirituality. It is a good day to help others, to ask for forgiveness, and to act with compassion. Conclude those unfinished projects and resolve any pending issues today. It is an auspicious day for women. It stands for realization.

Power 10 ~ *Lajuj* symbolizes the ten cosmic energies that relate to the fingers of the hands and the ten telluric energies of the toes. Take advantage of this energy to speed along your projects, activities, and aspirations. It can be a day of energetic highs and lows. It stands for intelligence, integrity, and balance.

Power 11 ~ *Junlajuj* symbolizes learning through experience. This can be a complicated day, as you are acquiring knowledge through experience, which can also bring trials. Be patient on this day and avoid making any important decisions. It stands for the cycles of life.

Power 12 ~ *Kab'lajuj* projects creativity, challenge, and dreams. Today you will have the energy to materialize your dreams. It is a good day to gain knowledge, increase your self-confidence, and help others. It is a day in which to practice sharing. It represents the consciousness of being.

Power 13 ~ *Oxlajuj* marks the end of the cycle and thus signifies achievement and success. Like Power 1 it contains strength and wisdom, but it has now gained experience. It is a day for rewards and spirituality; an excellent day to put your gifts to good use, to counsel others, and develop your intuition. It stands for fulfillment and magic.

The Fundaments of Maya Healing

Maya culture places a lot of importance on the integral development of the individual, which implies finding complete balance, both internally and externally. To accomplish this there must be stability between the four pillars that sustain our existence: the physical, mental, emotional, and spiritual planes. Inasmuch as we live in a reality that incessantly expresses polarity, we will always have to endeavor to achieve balance.

Healing, then, is a concept that involves energies; of balance between polarities and the planes of existence, as well as a return to what is natural, of respect for nature, and a life in harmony. This is why numerology and the Sacred *Cholq'ij* Calendar —as the axis of Maya civilization— play a fundamental role in their healing methods.

Among the Maya, the *Ajq'ij*, or spiritual guide that works as a healing specialist is the professional that treats all physical and emotional disturbances, and helps people consciously activate their inner energy. To this end, he avails himself to invocations and uses stones, incense, *"aguas floridas"* (perfumed water), and other ritual elements to unblock and balance the energetic centers of convergence in the body: the *Ukux* and the *Uxkanel*, described below, and to balance the polarities.

Numerology and the Cholq'ij in Healing

Two: *Opposing Energies*

The reality in which we abide bears polarity as one of its principles; it is a world of opposites: hot and cold, night and day, light and dark, feminine and masculine, among many others. One cannot exist, or find balance, without the other. We humans are not the exception to this rule, as we are made of positive and negative energy and posses a feminine and masculine side. Our left side expresses feminine, uneven-numbered energies, and our right side expresses masculine, even-numbered energies. Ailments arise when these polarities become unbalanced.

Right side of the body: masculine, even-numbered energies

Left side of the body: feminine, uneven-numbered energies

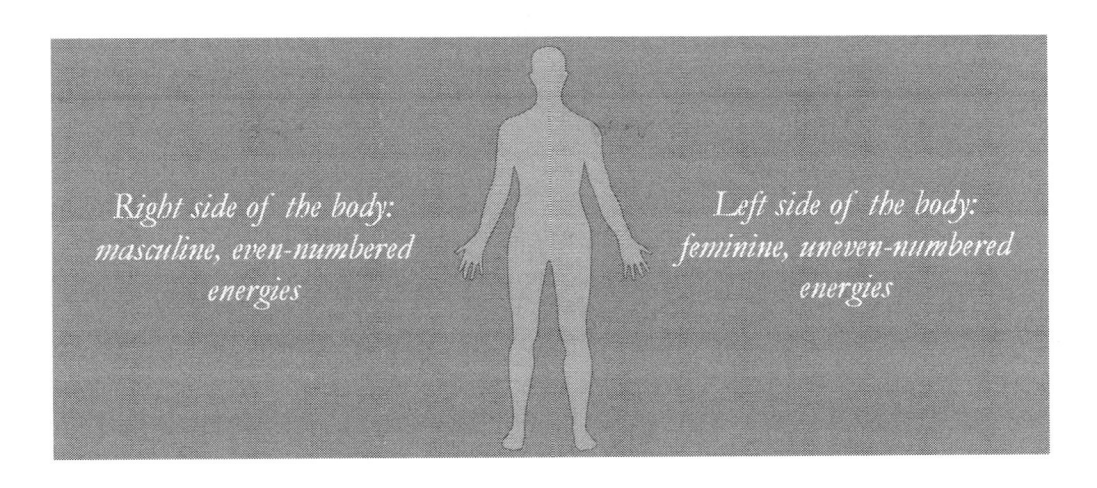

Four: *The Existential Planes*

As humans, we express ourselves in four planes of existence: the physical, mental, emotional, and spiritual planes. Whatever occurs in any of these planes will generally manifest in the physical plane, as this energetic unbalance will usually transfer to the physical body. Naturally, diseases caused by unhealthy eating habits, an inactive lifestyle, or accidents are not a reflection of the other planes but have a physical origin.

These four planes act as a unit. A good example is to imagine a table with four legs. If one of the legs is shorter than the others the table will be unstable; if one of the legs is weaker than the rest it will probably break. If the table is to be properly functional, all of its legs must be equally strong and have the same length. Health consists of searching for soundness in, and balance between, the four abovementioned planes. ¿How? By making an effort to achieve complete development in all planes; by dedicating ourselves to our physical and material needs without neglecting our intellectual, spiritual or emotional development. It is common for some people to develop only their material of physical aspects and forget about the rest, or to dedicate themselves to intellectual, spiritual, or emotional pursuits and leave the other aspects to one side —or not strengthen them sufficiently— and thus they become the "weak leg" of their table. In this way, it is the balance between both polarities, and the four planes of existence, that keeps us in perfect health.

Seven: *The Ukux*

According to Maya tradition there are 20 centers of energetic convergence in the physical body; seven of these are the *Ukux* or "hearts" and, although they are located in the physical body, they have an effect on the four planes. Consciously attaining the development of these centers is one of the purposes of the Maya tradition, as upon awakening them the true force of human beings becomes active and the dormant senses can then be developed.

Location of the seven Ukux

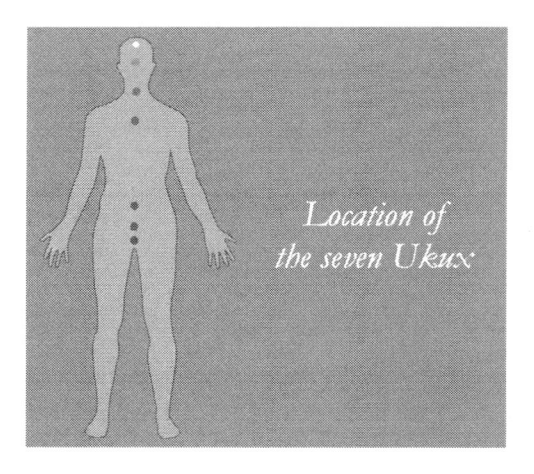

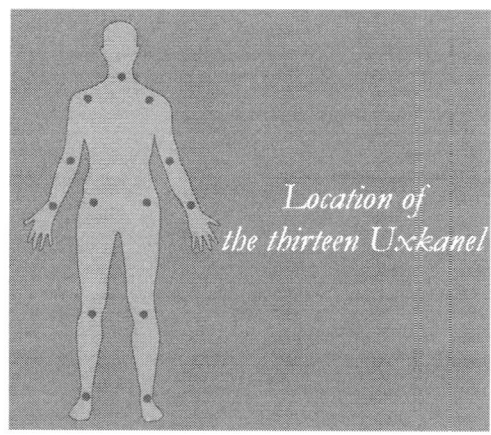

Location of the thirteen Uxkanel

Thirteen: *The Uxkanel*

After the seven *Ukux*, the other important 13 centers of energetic convergence are the *Uxkanel*, the joints. The *Uxkanel* are closely related to the *Ukux*. When energy flows from the *Uxkanel* the *Ukux* become activated, and vice versa. This allows energy to flow freely throughout our bodies. When there is a blockage in any of these points, energy will accumulate and affect the corresponding body part.

It is through the *Uxkanel* that our body can receive the cosmic and telluric energies, and these energetic entry points serve to activate and align our own energy. Another function of the *Uxkanel* is to manage the balance between our polarities.

Twenty: *The Cosmic and Telluric Energies, and the Integration of all Energies*

The reality in which we live is the result of the convergence of the cosmic (Heart of the Sky) and telluric (Heart of the Earth) energetic currents, which are related to the 10 fingers on our hands and the 10 toes on our feet, respectively. Our Grandfathers and Grandmothers say that we humans are like the Tree of Life because we have a connection with the underworld through our roots: our feet; we manifest ourselves in this world through our trunk, and have a connection with the heavens through our branches and crown: our arms and hands.

The 13 *Uxkanel* (joints) and the seven *Ukux* (hearts) are the 20 most important energetic entry points in our bodies, and it is on these that the *Ajq'ij* works with aromatic oils, stones, and the powers of the 20 *Nawals*. For this, he uses the Maya Cross of each person to locate the energy of the *Nawals* at each of these points. In this way, he can determine the illness that afflicts the person and balance his energy in a holistic manner.

Reincarnation in Maya Culture

Reincarnation in Maya culture is based on the principle of evolution, as there is no concept of involution. People are reborn at a higher vibrational level than they had in the previous life, until they reach the 13th heaven. At each of these heavens the human being must reach his full potential in that dimension to be able to pass to the next. Upon reaching the highest level he can become one with the Creator.

According to Maya culture humans are born to acquire learning, and this is achieved through the experiences encountered in the course of life. Not all obstacles and negative situations are seen necessarily as payment, but also as a way in which we learn and become strengthened, as it must be remembered that we are not here just to contemplate life: We are in this world to fulfill a purpose, and each of our daily actions are part of our evolution. This is the way in which we reach our true spiritual development. There is a law of cause and effect; however, when a person is going through a difficult situation it does not necessarily mean that he or she is paying for any negative actions in this life or past lives, as life is also meant to prepare us to become what destiny has marked for us, and this can also be achieved through adversity.

Toward the end of life, at the moment of our death, we must face ourselves, evaluate our actions without excuse or blame as to why we did not achieve certain things, and recognize the positive and negative actions we may have committed through our existence. After this analysis we can transcend to the dimension of death, from which we will reincarnate according to the vibrational characteristics in which we lived.

In addition to reincarnating in the ideal place and with the people we need to continue our development, we are also reborn with the *Nawal* and the Power that will afford us the personal traits that will help us reach the next spiritual level. For example, a person that constantly caused problems in the lives of those around him could be born under the *Nawal* 6 K'at, which is why he could face many setbacks throughout that life. On the other hand, a person with a weak character could reincarnate with the *Nawal* 13 Aj, which would grant him the strength of authority and success. On the contrary, a demanding person could be reborn under the *Nawal* and the Power in which his character would be manifestedly weak, which would allow him to learn and accept the opinions of others.

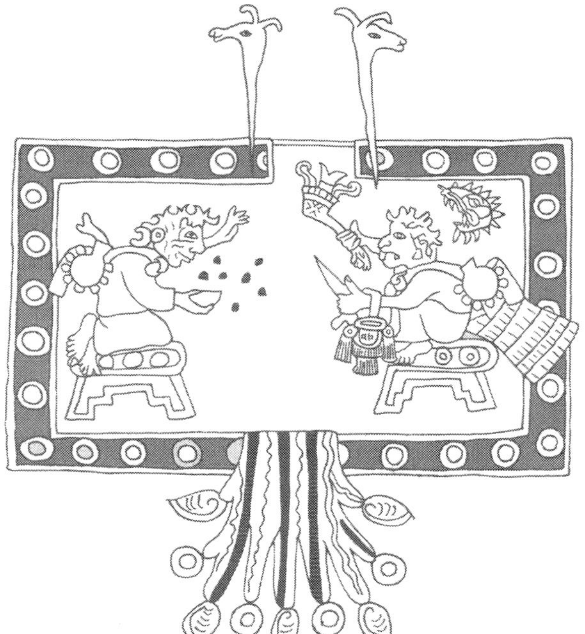

January 2012

Sunday	Monday	Tuesday	Wednesday	Thursday	Friday	Saturday
1 ☾ 13 Kan	2 1 Kame	3 2 Kej	4 3 Q'anil	5 4 Toj	6 5 Tz'i'	7 6 B'atz'
8 7 E	9 ○ 8 Aj	10 9 I'x	11 10 Tz'ikin	12 11 Ajmaq	13 12 No'j	14 13 Tijax
15 1 Kawoq	16 ☾ 2 Ajpu	17 3 Imox	18 4 Iq'	19 5 Aq'ab'al	20 6 K'at	21 7 Kan
22 8 Kame	23 ● 9 Kej	24 10 Q'anil	25 11 Toj	26 12 Tz'i'	27 13 B'atz'	28 1 E
29 2 Aj	30 3 I'x	31 ☾ 4 Tz'ikin				

JAN

January
Sunday 1

13 Kan

The Serpent of Fire breathes its mighty vigor into this day! Its convergence with Power 13 leads your projects to outstanding outcomes. You reach the end of this cycle alert, with renewed drive, and a profound sense of satisfaction.

January
Monday 2

1 Kame

You will finally transcend the earth-bound impasse you are in, as the energy of Power 1 grants you a fresh start. Connect with the Superior Beings and ask them for ancestral wisdom to guide your way, and for protection from any hardships during this cycle.

January
Tuesday 3

2 Kej

JAN

Although the day is charged with polarity, you will find strength and balance in the power that Kej has over the physical, mental, emotional, and spiritual manifestations of human existence, which is why you will be given the best that any opposing energies have to offer.

January
Wednesday 4

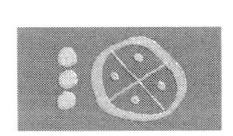

3 Q'anil

Anything that you start today will have a positive influence on your life. Connect with the Celestial Energies and ask the Creator to grant you the stamina you need to turn your dreams into reality, and the *Nawal* to bring you seeds of wisdom.

JAN

January
Thursday 5

4 Toj

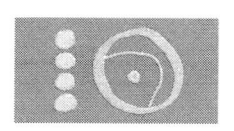

This is the day in which you are called to mend any errors you may have committed. Begin by offering a purple candle to the Creator and by asking the energies of the Power of Four to grant you resoluteness, so that from now on you may proceed in a more judicious manner.

January
Friday 6

5 Tz'i'

The energy of this day is infused with justice, order, and tolerance; use these values to encourage harmony in your community. Practice team work, as amazing things can be accomplished when everyone contributes!

January
Saturday 7

6 B'atz'

JAN

Today you have the initiative to carry out those ideas that have been lingering on your mind. However, the energy of Power 6 might hinder your intentions. Be strong, as you will have to fight to reach your goals.

January
Sunday 8

7 E

All roads embarked on under the auspices of these energies will grant you new knowledge and experiences that you can use for future projects. Tenacity is the clue to overcoming any difficulties that may arise, which in any case will help you grow into the person you really want to be.

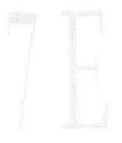

JAN

January
Monday 9

8 Aj

Aj, The Cane, represents abundance and endows you with core values and conviction. Together with Power 8, today it hands you all the elements you need to succeed in the financial arena. Remember: Sharing does not subtract, it multiplies!

January
Tuesday 10

9 I'x

The alliance of Power 9 with the feminine mystical powers of I'x projects a creative momentum that puts your mind into a receptive state and grants you the possibility of reaching the highest levels of consciousness. Use their energies to develop your psychic powers.

January
Wednesday 11

10 Tz'ikin

Today the energy that flows from Power 10 converges with Tz'ikin to put you into a positive and dynamic frame of mind. Take advantage of this great vibe; the *Nawal* of the day grants you its magnetic energy, so go on out and charm the world!

JAN

January
Thursday 12

11 Ajmaq

11 Ajmaq offers you the opportunity to undestand the significance of the choices you've made and to change discord into harmony. Today is a good day to undertake a full, unbiased inspection of your actions over the past 20 days; an exercise that will help renew your personal energy.

JAN

January
Friday 13

12 No'j

Light the flame that can shed clarity on your thoughts. Power 12 is the energy that activates your power of reflection, and its confluence with No'j transforms your knowledge into spiritual wisdom. Evolved spirits may gain access to other dimensions today.

January
Saturday 14

13 Tijax

Power 13 grants you the magic to transform everything that you want to change in your life, and to channel the energies in the precise direction that you wish. Use the double edge of Tijax to consummate a radical change; open up a path that inspires you!

January
Sunday 15

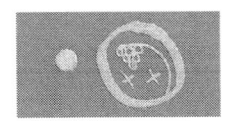

1 Kawoq

JAN

Kawoq gathers everyone in your community and offers the abundance that arises from fellowship, and Power 1 has the power and the strength of unity. Collective participation during this cycle will yield noble results.

January
Monday 16

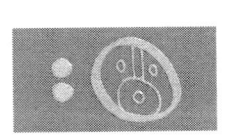

2 Ajpu

Positive and negative currents swirl randomly around you. Use the might of the Spiritual Warrior to overcome any tests that could arise during this day. Find a point of equilibrium between these energies and you will overcome any apprehension they may cause.

JAN

January
Tuesday 17

3 Imox

The left hemisphere of your brain is empowered with the convergence of Imox and Power 3. Ask the *Nawal* for the ability to invoke sacred messages through signs or dreams. This is a day in which unusual or unexpected things may occur.

January
Wednesday 18

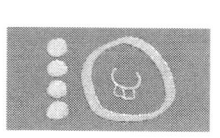

4 Iq'

Your mind is in a highly receptive and creative state today, which is why sudden, brilliant ideas may come to mind! What places or situations inspire you or stimulate your imagination? Power 4 helps you lay the foundation for a stable future.

January
Thursday 19

5 Aq'ab'al

This is truly a privileged day, as you will encounter a renewing force that will encourage you to embark in a new direction. However, you will have to overcome the duality that resides in Aq'ab'al to enjoy the rewards. Channel the *Nawal's* energies at dawn and dusk.

JAN

January
Friday 20

6 K'at

Look out! Complications may suddenly crop up today. Don't let yourself get caught up in the negative energies that permeate your surroundings. Try to maintain yourself centered and take things calmly to abate The Net's effects.

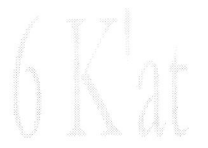

JAN

January
Saturday 21

7 Kan

Kan fortifies you today with its vital energy and places you in a balanced and harmonious state. Together with Power 7 it confers introspection and analysis so that you can put things into perspective, as today the collective memory of humankind is at your disposal.

January
Sunday 22

8 Kame

Ask the divine forces to shelter you from a bad death. It is a good day to light a candle and ask the Creator to withdraw any fatal diseases from your life and the life of your loved ones, and to petition for protection from accidents.

January
Monday 23

9 Kej

Kej exhibits the qualities that dwell in fire, earth, wind, and water. Center your thoughts on their energies to develop a closer connection with Creation, as Power 9 grants you the ability to understand Mother Nature's teachings.

JAN

January
Tuesday 24

10 Q'anil

Power 10 links you to the cosmic forces and the telluric currents. Let their energies flow through you so that the spiritual seeds of Q'anil can germinate, as they contain the codes that you need to conduct your life.

JAN

January
Wednesday 25

11 Toj

Keep the law of cause and effect in mind in everything you do today, as 11 Toj will put your capacity to learn —and your resilience— to the test. It is important that you take responsibility. Accept these trials to attain purification.

January
Thursday 26

12 Tz'i'

Today's energies grant a prompt solution to any disputes or injustices that may be occurring in your family or community, as truth will come to light and will guide everyone toward harmony and reconciliation. Any ill feelings will quickly be forgotten!

January
Friday 27

13 B'atz'

B'atz' is the beginning of infinite time, and its confluence with the experience and magical virtue of Power 13 gives you the elements you need to prepare original and creative plans that could fulfill your dreams in the near future!

JAN

January
Saturday 28

1 E

This cycle possesses the impetus of The Traveler and the new beginning granted by Power 1 to overcome any obstacle that may turn up along the way. It is a good day in which to conduct business abroad, as its energies open up many gateways!

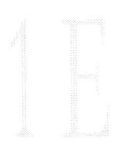

JAN

January
Sunday 29

2 Aj

Your partner is your biggest supporter, so join forces and work together! The energy that radiates from 2 Aj influences the accomplishment of shared goals, brings forth abundance, and lays down a solid base on which to build understanding in your home.

January
Monday 30

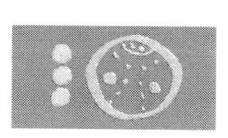

3 I'x

Channel the energy of I'x and allow its magic and intuition to guide you throughout the day, especially when managing occult forces. Use the astuteness it confers on you today to sharpen your vision and its strength to sustain you. Power 3 enhances the results.

February 2012

Sunday	Monday	Tuesday	Wednesday	Thursday	Friday	Saturday
			1 5 Ajmaq	2 6 No'j	3 7 Tijax	4 8 Kawoq
5 9 Ajpu	6 10 Imox	7 ○ 11 Iq'	8 12 Aq'ab'al	9 13 K'at	10 1 Kan	11 2 Kame
12 3 Kej	13 4 Q'anil	14 ☾ 5 Toj	15 6 Tz'i'	16 7 B'atz'	17 *Wayeb* 8 E	18 *Wayeb* 9 Aj
19 *Wayeb* 10 I'x	20 *Wayeb* 11 Tz'ikin	21 ● *Wayeb* 12 Ajmaq	22 *Ab'* **New Year** 13 No'j	23 1 Tijax	24 2 Kawoq	25 3 Ajpu
26 4 Imox	27 5 Iq'	28 6 Aq'ab'al	29 7 K'at			

January
Tuesday 31

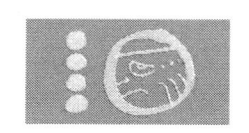

4 Tz'ikin

The energy projected by this day is charged with possibilities and experiences and predicts highly gratifying memories that will last you a lifetime, as Tz'ikin attracts material fortune, luck and love, and Power 4 provides its force and stability.

February
Wednesday 1

5 Ajmaq

Today you are granted spiritual strength to overlook the offenses you may receive from others. Make a conscientious decision to leave behind all resentment, and you will feel much happier, positive and at peace, as you will be totally liberated from these negative burdens.

February
Thursday 2

6 No'j

The convergence of these forces forebodes an agitated mental state and negative thoughts that could end in quarrels. Avoid any situation or social relationship that could exacerbate these tendencies. It is a good day to carry out your activities in solitude.

FEB

February
Friday 3

7 Tijax

The key word for today is "perseverance;" it is the only way you will get what you intend. The Knife will open a path and, although Power 7 could present some hurdles, it will equally allow you to advance to the position you have in mind.

FEB

February
Saturday 4

8 Kawoq

This day foretells financial wellbeing. Kawoq emanates abundance and plenitude, and Power 8 allows you to shape the conditions that will make it happen as it exercises all of its energy on your determination to achieve this goal.

February
Sunday 5

9 Ajpu

The energies of certainty and magic are your secret weapons on this day! Elevate your prayers to the Creator and Maker, as the elements you need to obtain spiritual realization can be attained with the vision granted by Ajpu.

February
Monday 6

10 Imox

Today's energy is somewhat ambivalent. You could have the strength and determination to achieve all your desires without difficulty, but things could also get complicated and slow you down.

February
Tuesday 7

11 Iq'

Focus your thoughts and ask Iq' —the wind that renovates— to help you cleanse your environment from all material and immaterial impurities. Ask the *Nawal* to send humankind the pure air of wisdom that teaches love and caring for Mother Nature.

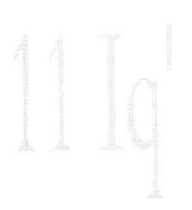

February
Wednesday 8

12 Aq'ab'al

The soft, peaceful energy brought by Power 12 and the grace of renewal granted by the *Nawal* that presides over the dawn allows you to retire all affliction from your life, so you can discover what is truly valuable.

February
Thursday 9

13 K'at

Take advantage of the energy of this day to shed light on the true nature of that conflict. K'at grants the mental ability to untangle the question, while Power 13 can help you wisely manage all of the negative consequences that it has heaped on you.

February
Friday 10

1 Kan

The level of energy of this new cycle is at a peak! Although Power 1 could cause some delay in your efforts to reach your goals with the desired speed, Kan provides you with so much power that you will reach them anyway.

FEB

February
Saturday 11

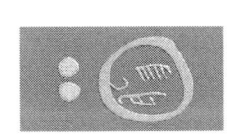

2 Kame

Kame will help you solve a negative situation, as it puts an end to any dispute that may have occurred as a result of two opposing views; but be careful how you manage its energies, as you run the risk of losing a valuable relationship in the process.

February
Sunday 12

3 Kej

Seek the stability you yearn for in nature and art. Surround yourself with these elements and you will be able to synchronize your energy with the Divine Energy that will bring you inspiring results and revelations, and will give you a feeling of peace and comfort.

February
Monday 13

4 Q'anil

Scatter many, many seeds of love during the course of your passage through this day! Without doubt, all the good actions you may plant today will bear positive fruits in your life and will steadily persist throughout the year.

February
Tuesday 14

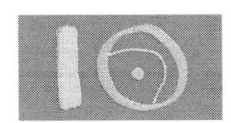

5 Toj

Thanks to the transcendence of Power 5 the energy of Toj restores all the positive actions that you have done over the past 20 days. Prepare to receive blessings and love in the same measure you may have dispensed them.

FEB

February
Wednesday 15

6 Tz'i'

Justice will prevail thanks to the energies of Tz'i', but certain facts may come to light that could bring conflict and confrontation; get legal counsel. It is in your best interest to take care of these matters promptly.

February
Thursday 16

7 B'atz'

This day might bring you many trials, but through them you will grow in understanding of yourself and gain wisdom. Think of these difficulties as the threads that add color, depth, and texture to the fabric of your existence.

February
Friday 17

8 E

E will guide your journey over the path that terrestrial forces have laid down for you today and will maintain you in close proximity with all things material. You will feel vibrant with energy, and might even transcend conventional limits if you make up your mind to do so!

February
Saturday 18

9 Aj

The magic of feminine energy will bring stability and abundance to your home on this day. Take into consideration the instinct and advice of women as today they are especially connected with superior forces. Your inner strength is fortified.

FEB

February
Sunday 19

10 I'x

Any obstacles in your path will easily fall by the wayside in the presence of your powers of mental concentration. The Jaguar's cunning, and the drive that Power 10 grants you today, will allow you to advance without difficulty toward the culmination of your ideas.

February
Monday 20

11 Tz'ikin

Be especially mindful that your material ambitions do not cloud your view and stop you from seeing things the way they really are. Make an effort to maintain yourself focused during this day, and analyze the possible consequences before acting.

February
Tuesday 21

12 Ajmaq

To find spiritual peace you will have to learn how to forgive the mistakes of others, but don't forget to forgive yourself! This day grants you the energy of forgiveness and the introspection that you need to achieve reconciliation.

February
Wednesday 22

13 No'j

No'j governs your intellect and endows you with knowledge and acumen, while Power 13 projects the energy of experience. Any pending plans or projects will report brilliant and creative results!

FEB

February
Thursday 23

1 Tijax

Don't wait any longer to make those changes you know you need to make. The energies of this day put the tools in your hands to achieve them, as Power 1 has the strength of new beginnings and the double edge of Tijax opens new paths.

February
Friday 24

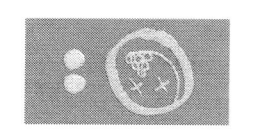

2 Kawoq

Friction with others is sometimes unavoidable but, luckily, the confluence of Kawoq with the union and support of Power 2 offers you the mechanism you need to mend any relationship that requires your attention.

February
Saturday 25

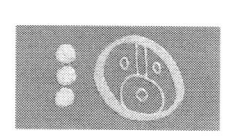

3 Ajpu

Take into account that whatever happens today is the result of your past actions. Ask the Creator for the courage to overcome your failures and for discernment to recognize the precise moment in which to go after any opportunities that may arise.

February
Sunday 26

4 Imox

Pay attention to the signs that may occur on this day, especially those that manifest in dreams; they will give you the clue you need to find balance. Light a blue, a pale-blue, and a white candle to activate your intuition and visions.

FEB

February
Monday 27

5 Iq'

Iq' brings with it inspiration; invigorate yourself with its energy, renew your spirit, and let it fill you with joy! Together with the energy of Power 5 it impels you to fly higher. Free yourself from any bonds, and let The Wind fill your sails!

February
Tuesday 28

6 Aq'ab'al

Use your inner strength and choose to have a positive attitude when facing life's problems. This may be a day of trials in which things could become complicated, but if you are prepared you can confront any difficulties with self-confidence.

February
Wednesday 29

7 K'at

Try not to be so analytical, as you could become mentally entangled in things that are really not that important. Keep things simple during this day, and you will have a clearer perspective of what may be happening around you.

March 2012

Sunday	Monday	Tuesday	Wednesday	Thursday	Friday	Saturday
				1 ☾ 8 Kan	2 9 Kame	3 10 Kej
4 11 Q'anil	5 12 Toj	6 13 Tz'i'	7 1 B'atz'	8 ○ 2 E	9 3 Aj	10 4 I'x
11 5 Tz'ikin	12 6 Ajmaq	13 7 No'j	14 8 Tijax	15 ☾ 9 Kawoq	16 10 Ajpu	17 11 Imox
18 12 Iq'	19 13 Aq'ab'al	20 **Spring Equinox** 1 K'at	21 2 Kan	22 ● 3 Kame	23 4 Kej	24 5 Q'anil
25 6 Toj	26 7 Tz'i'	27 **Waqxaqi' B'atz'** 8 B'atz'	28 9 E	29 10 Aj	30 ☾ 11 I'x	31 12 Tz'ikin

March
Thursday 1

8 Kan

Power 8 has the virtue of creating the energy that you need to carry out your plans. Today it increases its impact by combining with the intense and vibrant power of Kan. Together they grant you the ability to reach your highest levels of productivity.

March
Friday 2

9 Kame

Call upon the powers of Death to stop investing energy in things that leave nothing of value for your spiritual development. Power 9 is just the ally you need to help you clarify which are the situations that need reassessment.

March
Saturday 3

10 Kej

Opposing energies connect and feed off of each other during this day under the influence of Power 10, producing energetic highs and lows that are offset by Kej's intrinsic stability. Beware! It is a day in which you could suffer betrayal.

MAR

March
Sunday 4

11 Q'anil

Power 11 will bring everything that you plant on the day of the *Nawal* Q'anil to fruition. For this reason, and even if you are not asked to do so, you must put aside your self-interests and share your abilities and experience without expecting anything in return.

March
Monday 5

12 Toj

Power 12 displays its strength in actions that are undertaken collectively with one heart and one mind. Gather your family around the *Tojil*, the Sacred Fire, to elevate each person's petitions before the Creator and Maker.

March
Tuesday 6

13 Tz'i'

Hallowed Scriptures should be your main source of reference when walking the spiritual path. Tz'i' connects you with their sacred words, and Power 13 accords you the experience and wisdom to interpret their meaning so that you can apply their teachings in your daily life.

March
Wednesday 7

1 B'atz'

What are you choosing to record in your life history? Joy… fear… anger… compassion…? B'atz' —which contains the essence of your personal narration— and Power 1, shower their energies on you and offer you the possibility of reinventing yourself in this new cycle.

March
Thursday 8

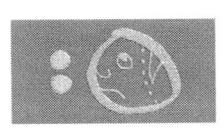

2 E

Power 2 fosters good work relationships or societies with people that could place you on the path to success, as E builds new avenues for you. It is an excellent day to set the wheels in motion for that plan you have in mind.

March
Friday 9
3 Aj

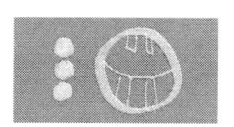

The fusion of the energies of Aj and Power 3 shines on everything related to your home today, especially on the children in your family. Offer the *Nawal* a white candle to ask for their wellbeing.

March
Saturday 10
4 I'x

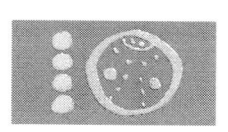

From within I'x gushes forth the creative force of Mother Earth. On this day it connects you with the force of the wisdom of the four *Balam'eb* —the men-gods— so that you can concern yourself with protecting Her creation and all sacred sites.

March
Sunday 11

5 Tz'ikin

Tz'ikin grants you the panoramic vision of the eagle that is free from the limitations dictated by time and space. Today you can rely on the energy of action that Power 5 provides you to tear down the barriers that separate you from your intentions.

MAR

March
Monday 12

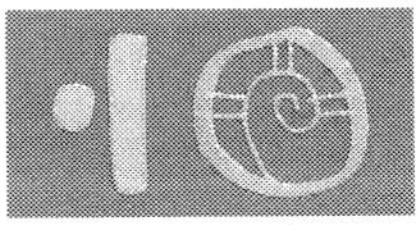

6 Ajmaq

Channel the energy of Power 6 to connect with your *Nawal* in the search for calm and prudence. Ask The Owl for wisdom in the management of the harmony and tension that this day's energies project, and for reconciliation with the Great Father.

March
Tuesday 13

7 No'j

Power 7 radiates intuition and inner growth by means of No'j, which governs your thoughts and connects your mind to the Universal Cosmic Mind. Open up to receive the gifts of wisdom and the knowledge that the *Ajaw* is granting you today.

March
Wednesday 14

8 Tijax

The powerful blade of Tijax allows you to cut out any negative energy, resentment, or disease that may have inadvertently entered your life, as you can also rely on the material force of Power 8 to help you carry out this task.

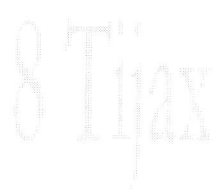

March
Thursday 15

9 Kawoq

The feminine force of Power 9 converges today with Kawoq, whose energy radiates on all matters concerning the family. Light a candle to ask for family union, blessings for your home, and protection for all.

March
Friday 16

10 Ajpu

Clarity in your actions will award you certainty and the strength to continue on the right path. Today you have the support of the Spiritual Warrior to conquer any harmful energies and the values of Power 10 to provide you with integrity.

March
Saturday 17

11 Imox

This day forebodes disquieting changes. Plan your next steps carefully, as Power 11 will probably put your decisions to the test. Guide your thoughts so that Imox cannot scatter your energies, and ask the *Nawal* to help you interpret any subtle manifestations.

March
Sunday 18

12 Iq'

Power 12 fuels your inner powers; light a white candle and meditate on its light so that it can connect you with the light of your inner fire. Iq' is the wind that will feed this fire with forcefulness, clarity, and sacred energy.

March
Monday 19

13 Aq'ab'al

The young and creative energy of Aq'ab'al unites with the learning and realization you have achieved through Power 13 during this trecena, and awards you faith and self-confidence. It is a day filled with well earned satisfactions that will let you break away from your routine.

March
Tuesday 20

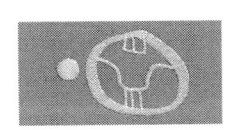

1 K'at

Use the forces of 1 K'at to link up with the cosmic and telluric energies of the **equinox**. Meditate on the internal fire in your coccyx until it courses up your column like the undulating light of *Kukulkán* and activates your internal powers. Today there is balance between polarities.

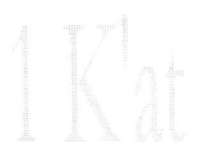

March
Wednesday 21

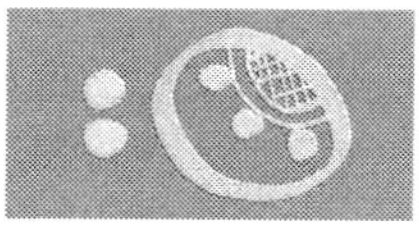

2 Kan

Kan possesses the sacred knowledge of sex, and it is the *Nawal* that activates sexual energy; Power 2 represents union and the couple. Today you will find balance and fulfillment in your sexual life, as you will attain energetic completeness with your partner.

March
Thursday 22

3 Kame

Hasty decisions can sabotage you, but today Kame allows you to return to the point of departure to rectify your mistakes. Keep your goals clearly in mind when asking Power 3 to connect you with the energies you need to undo any mix-ups.

March
Friday 23

4 Kej

Power 4 encompasses the physical, mental, emotional, and spiritual planes of being. Kej, The Deer, has a leg in each of the four cardinal points and sustains each of the four planes, and today you will find balance and a connection with everything!

MAR

4 Kej

March
Saturday 24

5 Q'anil

Q'anil symbolizes the seeds and the fruits of abundance; Power 5 represents work and opportunities. This is a good day in which to ask for the fruition of your projects, so petition with the knowledge that they will bring you personal satisfaction and financial wellbeing.

5 Q'anil

March
Sunday 25

6 Toj

The energies of 6 Toj project the law of cause and effect, which presages a difficult day. Carefully meditate your actions. You will need to build up a positive attitude for the events that might occur.

March
Monday 26

7 Tz'i'

The Sacred *Cholq'ij* Calendar reaches the end of its 260 cycle today. Light a white candle for each of the 20 *Nawals* that have accompanied you throughout this year as a sign of gratitude for the ancestral wisdom they have so richly bestowed on you through their energies.

March
Tuesday 27

8 B'atz'

Maya New Year The *Waqxaqi' B'atz'* New Year celebration marks the first day of the year on the Sacred Calendar. Its energies encompass all of the opportunities that this new phase will offer you so you can reach your ideals and accomplish the mission that life has chosen for you. Let us celebrate together around the Sacred Fire and ask the *Ajaw* —the Creator and Maker— for blessings, to thank Him for the opportunity that He grants us to rectify our path; for the opportunity of asking Him to *clarify* our path, and for the opportunity to begin anew!

MAR

March
Wednesday 28

9 E

This day possesses great positive energy. *Saq'be*, the White Road, connects you with the energies of the cosmos through the spirituality, love, and compassion of Power 9, and provides you the energy you need to perform to the best of your abilities.

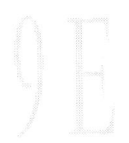

March
Thursday 29

10 Aj

Aj is the pillar that connects the cosmic and telluric energies of Power 10 with the Sacred Altar. Today you are in harmony with both divine and human laws. You may experience revelations in dreams.

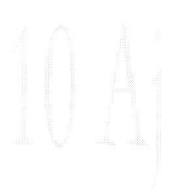

March
Friday 30

11 I'x

I'x emanates the power of the jungle and feline energy. Power 11 bids you to share and collaborate. Ask the *Ajaw* for the wellbeing of all living creatures, and send Mother Earth all the positive energy you can rally so that Her creative force may endure forever unaltered.

April 2012

Sunday	Monday	Tuesday	Wednesday	Thursday	Friday	Saturday
1 13 Ajmaq	2 1 No'j	3 2 Tijax	4 3 Kawoq	5 4 Ajpu	6 ○ 5 Imox	7 6 Iq'
8 7 Aq'ab'al	9 8 K'at	10 9 Kan	11 10 Kame	12 11 Kej	13 ☾ 12 Q'anil	14 13 Toj
15 1 Tz'i'	16 2 B'atz'	17 3 E	18 4 Aj	19 5 I'x	20 6 Tz'ikin	21 ● 7 Ajmaq
22 8 No'j	23 9 Tijax	24 10 Kawoq	25 11 Ajpu	26 12 Imox	27 13 Iq'	28 1 Aq'ab'al
29 ☾ 2 K'at	30 3 Kan					

March
Saturday 31

12 Tz'ikin

Shape your ambitions with the Power of 12, which brings creativity and is the energy that can materialize your wishes. You will accomplish anything you can image, as Tz'ikin gives you the tools you need to reach your goals.

APR

April
Sunday 1

13 Ajmaq

Ajmaq embodies the spirit of all great wise men and women; it is the *Nawal* that opens the lines of communication and allows you to absorb their teachings. Expand your awareness of your life purpose by heeding the messages of the Grandfathers and Grandmothers that will guide your way.

April
Monday 2

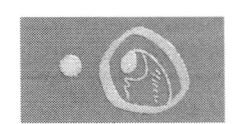

1 No'j

This cycle is under the aegis of No'j and will nurture your mind with knowledge, but first you must use what you already know and turn it into the wisdom you need for your personal growth. Take advantage of the impulse of Power 1 to make headway in that direction!

April
Tuesday 3

2 Tijax

Tijax allows you to cut out any negative social, love, or professional relationships to distance yourself from people with negative vibrations or attitudes of defeat. Once the scene is clear, new and stimulating changes will open up before you!

April
Wednesday 4

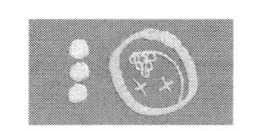

3 Kawoq

Power 3 encompasses the results of our actions and symbolizes our descendants. Its energy flows today with that of *Nawal* Kawoq's and strengthens family and community ties, and bids us to watch over their wellbeing.

April
Thursday 5

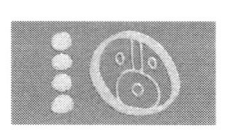

4 Ajpu

Ajpu is the force of life; it is the Father in His solar likeness and, you, as His child, can shine over those around you and send them your spiritual light. This day brings with it the strength of The Warrior and the stability of Power 4!

April
Friday 6

5 Imox

Imox offers you the gifts of intelligence, creativity and vigor; use them with the energy of love and joy granted today by Power 5 to celebrate life and everything it gives you! Connect with the Spirit of Water to explore your subconscious mind.

APR

April
Saturday 7

6 Iq'

Iq' is the element that rules over your thoughts, and governs change, communication, and the spoken and written word. Today it purifies your body, mind and spirit and brings beauty and harmony. Light a stick of incense to drive away the conflict and unbalance inherent in Power 6.

April
Sunday 8

7 Aq'ab'al

Aq'ab'al is the *Nawal* that awakens your consciousness; it is the new dawn that allows you to regenerate at every instant! From the energy of Power 7 emanates the impulse you need to get out of the rut and innovate.

APR

April
Monday 9

8 K'at

Power 8 contains the energy of art, creativity, and everything that relates with the material world. Establish a connection with your creative side to identify that unsatisfied desire to experiment and use your imagination without limits as, additionally, you have K'at to help you catch this wish!

April
Tuesday 10

9 Kan

This day offers you the space to perform in a productive environment that will render many benefits, as it is filled with the energy of realization that flows for Power 9, and Kan strengthens your inner fire. The Serpent will transmute negatives into positives.

APR

April
Wednesday 11

10 Kame

Kame closes another cycle and predicts changes. Today it converges with Power 10, which is the spiritual power that transfigures body and mind. Surrender to the forces of transformation, as the ancestors will accompany you on this journey!

April
Thursday 12

11 Kej

Power 11 represents the trials that you must undergo and the payments you must make to climb the steps of your spiritual evolution. Kej, which is balance and harmony, infuses you with its strength today so that you will not falter on this mission.

April
Friday 13

12 Q'anil

Power 12 alludes to the community and the consciousness of being. Sow the seeds of harmony and understanding in everyone you meet today with the help of Q'anil; doing this will bring much happiness into the lives of others!

April
Saturday 14

13 Toj

Toj is the special day that is reserved for making offerings. Show your gratitude to the *Ajaw* through the Sacred Fire for all that you have and will receive, and for everything that you are and have achieved, and ask for magic to come into your life.

APR

April
Sunday 15

1 Tz'i'

Tz'i' is a good day in which to ask for liberation from vices and poverty. Its energy intensifies with the influx of Power 1, which grants renewal and strengthens your resolve to begin a new chapter in your life.

April
Monday 16

2 B'atz'

Allow unconditional love to guide you on this day, and offer your partner your wholehearted support. Listen attentively without judging, giving your partner not only affection but also your time. With this, you will strengthen the ties that bind you.

APR

April
Tuesday 17

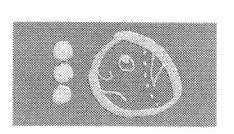

3 E

Power 3 will project the fruits of your actions. Ask the *Nawal* E —The Sacred Road— to guide you toward the path that can help fulfill your intentions, so that you can accomplish the mission for which you were born.

April
Wednesday 18

4 Aj

The solidity and conviction granted by Aj is complemented with the stability and strength provided by Power 4. The confluence of their energies will help you achieve a high degree of excellence and perfection in everything you do today.

APR

April
Thursday 19

5 I'x

Connect with the feline side of I'x and replenish yourself with the energy for action granted by Power 5. Feel the power conferred on you by Mother Earth and the Universe; their energies will flow through you and grant you wisdom far beyond your mental capabilities.

April
Friday 20

6 Tz'ikin

Ask the *Ajaw* for material fortune; The Eagle is the intermediary between heaven and earth and can make your wishes come true. Power 6 also represents the material world, but ask the Creator and Maker to guard you from its unstable energy.

April
Saturday 21

7 Ajmaq

Ajmaq embodies our mistakes and forgiveness. Ask the *Nawal* to help you recognize, and accept, the causes of your errors so you can be released from all blame. Upon converging with Power 7, its energies intensify, and it puts you on the path of inner harmony.

April
Sunday 22

8 No'j

The talent and mental acuity conferred on you by No'j today makes an important connection with Power 8, whose domain is the material world. This is a very fitting day to investigate and acquire knowledge on financial and business issues.

APR

April
Monday 23

9 Tijax

You can rely on the energy of The Obsidian Knife to banish the things that cause suffering in your life! Cut out all dependencies, addictions and fears, or whatever impinges on your development and happiness, and fill these spaces with the spiritual light granted by Power 9.

April
Tuesday 24

10 Kawoq

The Power of 10 will connect you with the prodigious forces of the cosmos! Ask the Supreme Energies to intercede on your behalf in matters concerning your social and work environments, especially regarding your bosses or coworkers, as Kawoq is watching out for you today.

April
Wednesday 25

11 Ajpu

Power 11 tends to diminish your level of vitality, but if you seek the support of Ajpu you can face any threat or challenge that comes your way today, as the *Nawal* will give you the wherewithal to leverage any unfavorable situation.

April
Thursday 26

12 Imox

Ask the *Nawal* Imox —who occupies the inner recesses of your mind— to banish any unproductive ideas and quiet your thoughts; you will find it much easier to cultivate your intuition and receive visions and messages in dreams.

APR

April
Friday 27

13 Iq'

Take a moment during the day to meditate. Concentrate on your breathing... drink in the pure and renovating air of Iq'... Allow it to turn your tension into relaxation, and to nourish your thoughts with every breath you take.

April
Saturday 28

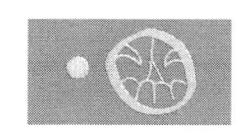

1 Aq'ab'al

Aq'ab'al grants you the clarity you need to recognize the opportunities that may arise today. Use its power to bring to light all that you believe is being kept from your knowledge. The energy of Power 1 allows you to accomplish your goals.

APR

April
Sunday 29

2 K'at

Stay alert! If you are uncertain about what you do or say today, K'at could take advantage of your indecisiveness and cause confusion and complications in your social affairs. On the contrary, if you focus, you could use this energy to catch desirable things for your personal relationships.

May 2012

Sunday	Monday	Tuesday	Wednesday	Thursday	Friday	Saturday
		1 4 Kame	2 5 Kej	3 6 Q'anil	4 7 Toj	5 8 Tz'i'
6 ○ 9 B'atz'	7 10 E	8 11 Aj	9 12 I'x	10 13 Tz'ikin	11 1 Ajmaq	12 2 No'j
13 ☾ 3 Tijax	14 4 Kawoq	15 5 Ajpu	16 6 Imox	17 **Rey San Pascual feast** 7 Iq'	18 8 Aq'ab'al	19 9 K'at
20 ● 10 Kan	21 11 Kame	22 12 Kej	23 13 Q'anil	24 1 Toj	25 2 Tz'i'	26 3 B'atz'
27 4 E	28 ☽ 5 Aj	29 6 I'x	30 7 Tz'ikin	31 8 Ajmaq		

April
Monday 30

3 Kan

Capture the *joie de vivre* and vitality of the children that surround you! Kan increases your physical strength and grants you the untiring energy that gushes from their youth. Power 3 offers you as well the gift of imagination and creativity.

MAY

May
Tuesday 1

4 Kame

This is a good day to ask for spiritual guides to enter into your life, as well as for Grandfathers and Grandmothers that can transmit their ancestral wisdom and point you in the direction you need to take in your search for spiritual enlightenment.

May
Wednesday 2

5 Kej

Work with joy and enthusiasm! Your relationship with the four elements and with Mother Earth is privileged today thanks to the influence of Kej. Outwardly express your happiness in everything you do, as Power 5 will bring you many rewards.

MAY

May
Thursday 3

6 Q'anil

The Seed supports your attempt to change those things that you would like to improve in yourself, and Power 6 helps you polish your personality. It is a good day to ask the energies to cast out timidity or arrogance from your life.

May
Friday 4

7 Toj

Thank Father Sun for shining on your life every day. It is a good day to pay for all that has been granted you, good or bad, and to make an offering for the things that you have not yet received. The Sacred Fire opens the channels of communication with the Divine Energies.

MAY

May
Saturday 5

8 Tz'i'

The material justice personified in Tz'i', and the material authority represented by Power 8, favor the resolution of any legal or financial issues on this day. Light a candle to the *Nawal* to ask for deliverance from poverty.

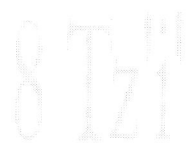

May
Sunday 6

9 B'atz'

Feminine energies prevail in 9 B'atz'. This is a day in which to celebrate the maternal womb called Planet Earth, which gives us life and offers us everything we need to satisfy our physical and emotional needs. Let her unequaled beauty inspire you!

MAY

May
Monday 7

10 E

E symbolizes your destiny; the path of existence that guides you toward your purpose in life and in which you will find realization. Ask Power 10 to connect you to the telluric energies so that they may guide your footsteps.

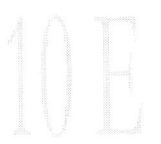

May
Tuesday 8

11 Aj

Apply everything that you have learnt through the experiences that Power 11 has provided; the influence of Aj grants you the admiration of everyone around you and allows you to emerge as a leader in any activity you take part in today.

May
Wednesday 9

12 I'x

Hidden forces bow before the power of I'x, which has an ally in Power 12 today. Their energies increase your self-confidence and allow you to face any unknown situation by granting you discernment and the ability to expand your mind.

May
Thursday 10

13 Tz'ikin

Ask The Eagle to create the ideal conditions for true love to enter into your life, or for soul mates in whose friendship you can find solace and loyalty. Power 13 has magic and will make your wishes come true!

MAY

May
Friday 11

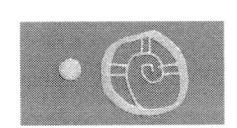

1 Ajmaq

Power 1 contains absolute knowledge, and Ajmaq is the curious mind that is receptive to constant learning. Express your beliefs but learn to listen to others; in this way you will find the wisdom that leads to genuine transformation.

May
Saturday 12

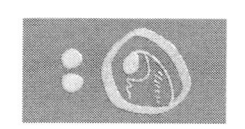

2 No'j

The balance between the positive and negative energies in Power 2 is influenced today by No'j, which brings harmony and clarity. This is a good day to unite your goals with the goals of others, and to give and receive support!

MAY

May
Sunday 13

3 Tijax

To achieve the positive results that Power 3 can offer you today, you must use the sharp edge of Tijax to cut out all that mental baggage that you have been dragging along from your past. Take advantage of this day's energies to give yourself a physical and energetic cleansing.

May
Monday 14

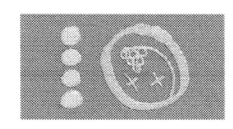

4 Kawoq

This is a good day to start or search for a new job. With the help of Power 4, there will be stability in the occupation you've chosen, and The Turtle will bring unity, abundance, and blessings to those whose activities promote the wellbeing of others.

MAY

May
Tuesday 15

5 Ajpu

Power 5 projects creativity, action and energy, and Ajpu bears art and magic. Take advantage of their combined values to stretch the limits you have unwittingly imposed on your creative faculties, and you will be pleasantly rewarded with the results. Dance with The Warrior!

May
Wednesday 16

6 Imox

6 Imox could be a strange and contradictory day as it forecasts trials, unbalance, and confrontation. Use your intuition to manage these vexing situations. Meditation could help maintain your mind in a calmer state to analyze the circumstances before reacting.

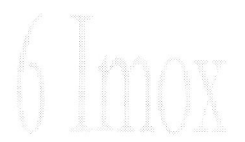

MAY

May
Thursday 17

7 Iq'

Iq' is the perfect day to renovate. Breathe deeply and recharge your strength with the fresh air that will nourish your body, soul, and spirit. Power 7 brings harmony and balance. Cast your cares to the wind!

May
Friday 18

8 Aq'ab'al

This is a good day to start a business or find a new home. Aq'ab'al opens the doors to the opportunities that are coming your way, and shines a light on the decisions you need to make. Power 8 you can grant you material stability.

MAY

May
Saturday 19

9 K'at

K'at's *Nawal*, The Iguana, and Power 9, symbolize fertility and the fruit, making this a good day for couples that wish to have children to petition the *Ajaw* to grant them the wondrous joy and blessings of parenthood!

May
Sunday 20

10 Kan

Kan helps you manage any anger that you may be feeling toward people or situations that have aggrieved you. Ask the energies of the day to grant you serenity, and the *Nawal* to show you how to manage this emotion and transform it into forgiveness.

MAY

May
Monday 21

11 Kame

Kame portends transformation and Power 11 brings trials that contain many lessons. Stop attempting to cling to people, ideas, or things, because today you will come to the realization of what it is that you must put behind you. Life is constant change, so practice letting go.

May
Tuesday 22

12 Kej

Kej travels on the four roads toward the four cardinal points; gather the power that The Deer will bring for you from each of the four corners of the world! Ask Power 12 to grant you protection against treason.

MAY

May
Wednesday 23

13 Q'anil

There is a time to plant the seed and a time to harvest the crop. Power 13 indicates that the moment has come for you to gather the fruits or your efforts, but be sure to plant new seeds in fertile ground for the coming stages!

May
Thursday 24

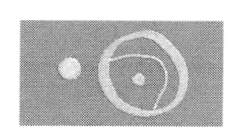

1 Toj

Toj is a day for offerings; a day in which to calm any instability that may be manifesting in your life, and a day that brings justice and hope. It is a good day to ask to be paid what you deserve for your work!

MAY

May
Friday 25

2 Tz'i'

Today's energies project clarity and authority; thus the truth will be revealed. Settle any pending issues. The *Nawal* and the Tone will exercise their justice in any difficult situation in which you may be involved.

May
Saturday 26

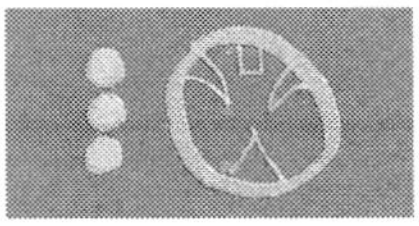

3 B'atz'

B'atz' is the thread that weaves your destiny and the umbilical cord that connects you to your lineage through your mother's womb. Use the energies of this day to strengthen your family ties and to settle any disagreement that may exist among you.

MAY

May
Sunday 27

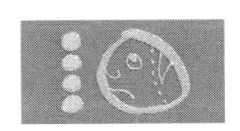

4 E

This is a good day to plan or start a trip, as the *Nawal* E opens up the path and protects your footsteps. Power 4 grants you the physical, mental, emotional, and spiritual balance to embark on an exciting adventure!

May
Monday 28

5 Aj

Aj is The Pillar that connects the cosmos with the earth and activates your vitality. Embrace a tree; it will put you in contact with the energy of Mother Earth through its roots, and with the cosmos through its crown!

MAY

May
Tuesday 29

6 I'x

Your day could become complicated, especially in situations that may involve women. Try to avoid crowds and disputes, as the low, confusing energies that prevail today could bring uneasiness and irritation. Enjoy your own company!

May
Wednesday 30

7 Tz'ikin

Tz'ikin is a special day to develop extrasensory abilities, especially revelations in dreams, vision, and intuition. Additionally, Power 7 increases your ability to acquire special powers. Activate these extraordinary tools!

MAY

May
Thursday 31

8 Ajmaq

Today your ancestors offer you the best of their experience and wisdom. Feel the profound and indelible connection between you, and ask them to guide you and send you the answers that you need. Mitigate your mistakes through community service.

Junio
Friday 1

9 No'j

This day is a day that grants spiritual knowing. Power 9 connects you with your mystical side and with No'j, which governs the mind. Today you are granted the ability to attract healing energies. Light nine white candles.

Junio
Saturday 2

10 Tijax

Let Tijax liberate you from the ties that are subconsciously binding you to any physical or moral suffering, and use the restoring energies of Power 10 to renew your connection with Heart of the Sky and Heart of the Earth!

June 2012

Sunday	Monday	Tuesday	Wednesday	Thursday	Friday	Saturday
					1 9 No'j	2 10 Tijax
3 11 Kawoq	4 ○ 12 Ajpu	5 13 Imox	6 1 Iq'	7 2 Aq'ab'al	8 3 K'at	9 4 Kan
10 5 Kame	11 ☾ 6 Kej	12 7 Q'anil	13 8 Toj	14 9 Tz'i'	15 10 B'atz'	16 11 E
17 12 Aj	18 13 I'x	19 ● 1 Tz'ikin	20 **Summer Solstice** 2 Ajmaq	21 3 No'j	22 4 Tijax	23 5 Kawoq
24 6 Ajpu	25 7 Imox	26 8 Iq'	27 ☾ 9 Aq'ab'al	28 10 K'at	29 11 Kan	30 12 Kame

June
Sunday 3

11 Kawoq

Power 11 reveals what you have experienced and learned. Kawoq channels the wisdom of the home and upholds the rules that families need to live in harmony. Show everyone that you value their contribution: Allow them to express their ideas and opinions without judging them!

JUN

June
Monday 4

12 Ajpu

This is a good day to ask The Hunter for analytical abilities and concentration to focus on your plans. The *Nawal* will grant you the certainty you need, and Power 12 will provide the energy to materialize your dreams.

June
Tuesday 5

13 Imox

Imox brings you magic and the power to transform any situation, no matter how unfavorable it appears at this time! Proceed with the wisdom granted by Power 13 so that these changes are fair for all, and beneficial for your future.

June
Wednesday 6

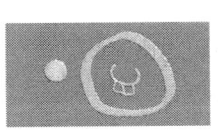

1 Iq'

Let Power 1 lift your prayers to the Creator and Maker with the force of the wind. Invite the activating energies of Iq' to spur you into action, so that you can work on accomplishing your goals.

JUN

June
Thursday 7

2 Aq'ab'al

Aq'ab'al is light and shadow, strength and weakness... its energies are antipodal yet harmonious. Likewise, Power 2 contains polarity and balance. Although you must walk today among opposing energies, your clear-sightedness will allow you to choose the right alternatives without fear of making any mistakes!

June
Friday 8

3 K'at

Today you will have the help of K'at to resolve those emotional or love-related problems that are weighing you down. The Net, in convergence with Power 3, will not only offer you surprising solutions, they will also be lasting.

June
Saturday 9

4 Kan

Your physical, spiritual, and sexual energies are in harmony today, as Kan's mighty vigor is connected to Power 4's balancing forces. Light your inner fire and feel its energy recharge you!

June
Sunday 10

5 Kame

Kame projects the energy of transmutation. Petition the *Nawal* to guard you during your journey, harmonize any negative energies, and to protect you from a bad death, as Power 5 will do its part to mitigate any unwanted outcome.

June
Monday 11

6 Kej

Beware of treason; today it can come from where you least expect it! Kej will protect you from those that wish you harm and from the conflicts projected by Power 6, and guides your destiny at dawn in the East, at nightfall in the West, on the Path of Water in the South, and on the Windy Path in the North.

JUN

June
Tuesday 12

7 Q'anil

Spiritual seeds flourish on the day of Q'anil; it is a good day for self-reflection. Power 7 offers you the faith, intuition, and introspection you need for inner growth. Sow kindness and compassion and you will harvest love and harmony!

June
Wednesday 13

8 Toj

Power 8 grants you the authority and moral strength to engage in your spiritual struggle. Enlightened beings whose mission it is to lead you toward the light gather under the energy of Toj. Don't miss this transcendental *rendezvous* with your guides!

JUN

June
Thursday 14

9 Tz'i'

At this emblematic moment, Tz'i' converges with Power 9 and increases your level of connection with all that is intuitive, creative and artistic. It is a good day to pick up a pen and write down your most intimate reflections on paper.

June
Friday 15

10 B'atz'

Power 10 fulfills all of your desires! Channel its energies to communicate with the Superior Energies and ask them, by means of B'atz', to help you find that special person that will satisfy and complement you in every way.

JUN

June
Saturday 16

11 E

The energetic highs and lows emitted by Power 11 today will stop you from making any important decisions. To avoid any further setbacks, light a candle to the *Nawal* and ask to be pointed in a hitherto overlooked direction, and for a path that is free from obstacles.

June
Sunday 17

12 Aj

From The Scepter radiates a worldly power whose energy rises like a column that sustains your inner strength, your principles, and your virtues. Power 12 wields its influence over you today and increases your self-confidence, and your prestige and authority in the eyes of others.

June
Monday 18

13 I'x

I'x endows you with the power to expand your mind and connects you with the highest levels of consciousness. Use this knowledge, and the maturity you've gained during this cycle, to overcome the forces that may be delaying your evolution.

June
Tuesday 19

1 Tz'ikin

Your financial success is stimulated today by Tz'ikin, whose energy grants you a futuristic and creative view that allows you to crown your ambitions regarding these matters, and you have the energy of Power 1 to perfect your strategies!

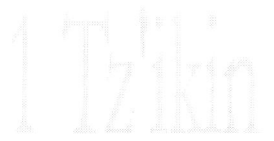

June
Wednesday 20

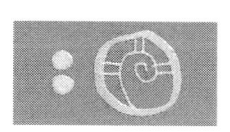

2 Ajmaq

Today your body is more sensitive and open to receiving the energies of the **solstice**. Meditate to connect with the sun at noon… feel how it recharges you… Open up to the wisdom the Elders are sending you through Ajmaq!

June
Thursday 21

3 No'j

3 No'j grants you the gift of communication and has a positive influence on your relationship with children and youth. Share your feelings and advice with patience and humbleness; today you speak their language, and they are eager to listen!

June
Friday 22

4 Tijax

Peer deeply into the black, polished blade of Tijax to enter into the dimensions of thunder and lightning. Grip it strongly and tear open the veil of mysteries. The wisdom of the four *Balam'eb* will accompany you to guide your hand.

June
Saturday 23

5 Kawoq

The energies of prosperity and abundance knock on your door today; open it up to receive them! Fill a small, green pouch with sesame seeds and put it in your wallet. Early in the morning you can also sprinkle some sugar at the entrance of your place of business or work.

June
Sunday 24

6 Ajpu

Pay attention to the events that transpire around you today, so you can recognize the right moment in which to make the move toward which Power 6 is thrusting you. Ajpu commands your marksmanship, so you don't miss the shot!

June
Monday 25

7 Imox

Enter the recesses of your mind with the power of Imox, who will connect your reflections like streams that become mighty rivers on their way to the ocean. In addition, Power 7 grants you introspection and the power of analysis that leads to clarity.

JUN

June
Tuesday 26

8 Iq'

The Wind and Power 8 grant all men the power of authority today, and a whirlwind of vital energy! Excellent financial opportunities are forecasted by their energies, which also bring forth beauty and harmony.

June
Wednesday 27

9 Aq'ab'al

A project that is left unfinished contains wasted energy that, upon lacking a goal, can slow you down. However, Aq'ab'al can offer you solutions, and Power 9 has the impetus to help you finish it and lift this weight off your shoulders.

June
Thursday 28

10 K'at

Use the cosmic energies that flow through your fingers today to untie any physical, mental, or spiritual oppression in which The Net may have trapped you, and to fling it again into the Universe to capture the new life experiences that await you!

June
Friday 29

11 Kan

11 Kan projects the changes you undergo within as time goes by. This day offers you the opportunity to examine the intrinsic value of your beliefs. Which should you preserve? Which are best discarded?

June
Saturday 30

12 Kame

Put an end to your self-centeredness and isolation. Find happiness through mentoring and sharing your talents with those around you. Your contribution to the community is very valuable and far-reaching!

JUN

July
Sunday 1
13 Kej

Kej will help you maintain your equanimity. The degree maturity you have reached at the closing of this cycle will provide you a new perspective. Stop worrying about a past that cannot be changed, and busy yourself with preparing for the next level.

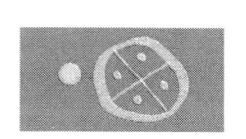

JUL

July
Monday 2
1 Q'anil

Embrace the mentality of a pupil, because this new cycle will bring you many lessons! Power 1 has the energy of new beginnings: Open your mind so that Q'anil can firmly plant within your psyche the kernels of reflection that lead to profound transformation.

July 2012

Sunday	Monday	Tuesday	Wednesday	Thursday	Friday	Saturday
1 13 Kej	2 1 Q'anil	3 ○ 2 Toj	4 3 Tz'i'	5 4 B'atz'	6 5 E	7 6 Aj
8 7 I'x	9 8 Tz'ikin	10 9 Ajmaq	11 ☾ 10 No'j	12 11 Tijax	13 12 Kawoq	14 13 Ajpu
15 1 Imox	16 2 Iq'	17 3 Aq'ab'al	18 4 K'at	19 ● 5 Kan	20 6 Kame	21 7 Kej
22 8 Q'anil	23 9 Toj	24 10 Tz'i'	25 11 B'atz'	26 ☽ 12 E	27 13 Aj	28 1 I'x
29 2 Tz'ikin	30 3 Ajmaq	31 4 No'j				

July
Tuesday 3

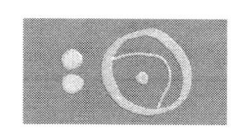

2 Toj

The energies of duality, dichotomy, and contradiction are intensely powerful today. Things may not be as they appear! All action evinces a reaction; if you are surprised by the repercussions it is because you are acting without thinking.

JUL

July
Wednesday 4

3 T'zi'

The Dog is the faithful guardian of Divine and Natural Law, and Power 3 materializes your actions. This is a day of spiritual and material justice. Ask the *Nawal* to liberate you from malicious criticism and rumors.

July
Thursday 5

4 B'atz'

B'atz' is the beginning of time without end. It is the thread of continuity between your past, present, and future. Resort to the energy of Power 4, and your present will always be stable! It is a good day for marriage.

July
Friday 6

5 E

JUL

The energies of this day bring you cheer and camaraderie thanks to the magnetism and charisma of Power 5, and E will guide your emotions and sentimental affairs. Have fun!

July
Saturday 7

6 Aj

Disagreements that erupt in discussions, and changes in the responsibilities of family members tend to occur under these energies. However, Power 6 also entails learning, so the pillars of the home will be strengthened by the end of the day.

July
Sunday 8

7 I'x

Your sense of perception will be heightened today; open your sensibilities to Mother Earth's subtle signals. Power 7 confers intuition and I'x grants you the gift of penetrating vision. Walk the path of the The Jaguar…

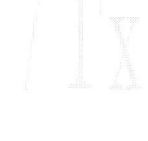

July
Monday 9

8 Tz'ikin

Your financial prosperity could well be at its peak! The magnetic energy of Tz'ikin attracts abundance, luck, and success, and the material energy of Power 8 predicts substantial earnings. Don't forget to save for a rainy day!

July
Tuesday 10

9 Ajmaq

JUL

Feel how Power 9 envelops you in its warm and loving vibrations! It urges you to get in contact with your emotions and act with compassion. Call upon the energy of Ajmaq, so that the words of the Elders dwell within you.

July
Wednesday 11

10 No'j

Take advantage today of the energies of Power 10 that bind all earthly and celestial intentions, and help you get your plans underway. No'j connects you with the Universal Mind, which is pure and omniscient intelligence.

July
Thursday 12

11 Tijax

The energy of 11 Tijax harbors the telluric force of earthquakes and has the power to bring about great changes. This could be a convulsive day, filled with setbacks and aggravations. Act cautiously and ask the *Nawal* for discernment through the use of crystals.

July
Friday 13

12 Kawoq

The energies of this day extol awareness and the power of shared wisdom. The community meets under the auspices of Kawoq so that we may selflessly participate, share, and support one another. Share your knowledge, abilities, and experience!

July
Saturday 14

13 Ajpu

JUL

You finally hit the bullseye! Sure enough, you've been practicing for days, and after a number of missed shots, The Blowgun Hunter makes an appearance and helps you in your efforts. The reward is that you can ask for whatever you want, because today you will get it.

July
Sunday 15

1 Imox

This day could flow with some difficulty, as Power 1 does not yet have the experience to channel the fluctuating energies of Imox, although it will provide you strength and perseverance to continue ahead. It is a good day to readjust your plans.

July
Monday 16

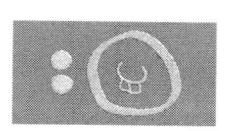

2 Iq'

You could feel somewhat unsure when making a decision today because Power 2 projects both balance and instability. Allow Iq' —the *Nawal* that governs the mind— to carry you effortlessly to new heights; you will have a better vantage point when considering the options.

July
Tuesday 17

3 Aq'ab'al

Aq'ab'al has gathered a fistful of marvelous rays of light for you today that will help dissipate your tedium. Power 3 infuses you with creative energy and optimism. Lift your spirits and get yourself out and about again!

July
Wednesday 18

4 K'at

Cast The Net high and wide, and catch anything you wish! Power 4 predicts that everything you gather today will endure and will contribute toward creating a stable life for you. Concentrate on the power of positive thinking!

July
Thursday 19

5 Kan

Become a person of action! Give it your best and seek to conquer the highest peak in everything you do. Acknowledge the things that come into your life and accept that they happen for your own good. 5 Kan grants you the energy to go farther.

JUL

July
Friday 20

6 Kame

Avoid any activities that could endanger your physical wellbeing. Light a black candle and ask the Creator and Maker to drive away the bad energies that maraud on this day, and for protection from fortuitous events.

July
Saturday 21

7 Kej

Find a moment to dedicate to yourself for personal renewal. Nature calls on you today to recharge your energy with its songs, aromas, and peace. It exists for your wellbeing and pleasure: Receive these gifts with gratitude!

July
Sunday 22

8 Q'anil

JUL

From 8 Q'anil springs forth the energy of creation. Today you can ask for a good planting and a good harvest; for the protection of crops, and for the recovery of sterile ground. This is the day in which to celebrate mothers with child!

July
Monday 23

9 Toj

Connect with your inner light through the energy of Father Sun. Ask the *Ajaw* that your gifts may be wakened through the energy of the Ceremonial Fire, and Power 9 to make you sensitive to the needs of others. Shine your light and compassion on everyone around you!

July
Tuesday 24

10 Tz'i'

Tz'i' wields the power of the sacred writings whose vibrations shed light upon the mind. On this day, you will enjoy spiritual and material balance. Your creative side can receive the cosmic energies that Power 10 radiates through the written word.

July
Wednesday 25

11 B'atz'

Power 11 projects learning through experience. You will need the strength of the *Nawal* to maintain your focus. What is time worth? Make each moment count and you will understand; whatever it is that you wish to accomplish in life depends on this.

July
Thursday 26

12 E

JUL

Have you found your way? Power 12 indicates that your dreams will come true. Are you still searching? Power 12 will give you what you need to acquire knowledge and increase your self-confidence. In either case, pay attention to the signs that may unexpectedly appear during the day.

July
Friday 27

13 Aj

Aj raises the pillars of inner strength and determination to sustain you; count on the support of the *Nawal* to accomplish your goals! Thanks to Power 13 you will experience a renewed and profound sense of realization.

July
Saturday 28

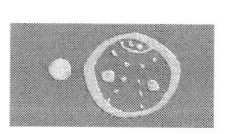

1 I'x

Activate the strength, cunning, and power that I'x grants you today to reach the highest levels of consciousness. Power 1 gently urges you to accept the responsibility of guiding your life in a more creative way, and grants you the strategies to achieve it.

July
Sunday 29

2 Tz'ikin

Don't resign yourself to live without love. Do what you love, and learn to love what you do. Love everyone and love yourself. Power 2 proclaims union and support; ask Tz'ikin to make sure that love is present in every moment of your life!

July
Monday 30

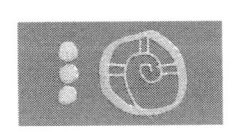

3 Ajmaq

JUL

Unconditional forgiveness will free you from affliction and the heavy energies of resentment. Ask the *Nawal* to help you transform any hard feelings into inner peace, so that you can overcome this obstacle in your quest for spiritual growth. Power 3 infuses you with the will to achieve it.

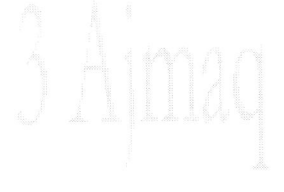

July
Tuesday 31

4 No'j

You can materialize anything you can create with your imagination, because No'j emanates the power that blurs the rigid limits of reality and allows your mind to transcend without obstacles to other planes, and Power 4 puts you in harmony with the four manifestations of your human nature.

August
Wednesday 1

5 Tijax

5 Tijax is a magical day! Power 5 opens the doors to the intangible, and the *Nawal* possesses the force of lightning that awakens visions. Get charged with the psychic and magical energies that infuse this day, so that your intentions come true according to your desire.

August 2012

Sunday	Monday	Tuesday	Wednesday	Thursday	Friday	Saturday
			1	2 ○	3	4
			5 Tijax	6 Kawoq	7 Ajpu	8 Imox
5	6	7	8	9 ☾	10	11
9 Iq'	10 Aq'ab'al	11 K'at	12 Kan	13 Kame	1 Kej	2 Q'anil
12	13	14	15	16	17 ●	18
3 Toj	4 Tz'i'	5 B'atz'	6 E	7 Aj	8 I'x	9 Tz'ikin
19	20	21	22	23	24 ☽	25
10 Ajmaq	11. No'j	12 Tijax	13 Kawoq	1 Ajpu	2 Imox	3 Iq'
26	27	28	29	30	31 ○	
4 Aq'ab'al	5 K'at	6 Kan	7 Kame	8 Kej	9 Q'anil	

August
Thursday 2

6 Kawoq

The energy of this day displays the conditions that could influence the appearance of conflicts in the family. Knowing this ahead of time will help you proceed with caution and wisdom. Words are very powerful; refrain yourself from speaking without thinking. Say what you think and feel what you say, but with prudence.

August
Friday 3

7 Ajpu

Conscientiously acknowledge the spiritual light that Father Sun is sending you today through Ajpu. Power 7 grants introspection and inner calm. It is a mystical and sublime day in which good can triumph over evil.

August
Saturday 4

8 Imox

Imox grants you the sweeping force of water and channels your actions toward achievable goals, and the physical energies of Power 8 will to see to it that your efforts materialize. Only those plans that have been diligently deliberated will render benefits.

August
Sunday 5

9 Iq'

The Wind's soft breeze penetrates your consciousness, blowing away old notions and opening spaces for new ideas. Thanks to the wise energy radiated by Power 9 you can formulate the questions to which your subconscious will provide the answers.

AUG

August
Monday 6

10 Aq'ab'al

Unfasten your moorings at the break of dawn and focus your eyes on a new horizon, as Aq'ab'al will lead you to safe harbor by nightfall. Breathe easily; Power 10 will see to it that the day fulfills your desires.

August
Tuesday 7

11 K'at

Try to accept the problems that come into your life as serenely as possible, as they are the anvil that strengthens your character and sharpens your resourcefulness. Leave any important decisions for another day.

August
Wednesday 8

12 Kan

Light a red candle to connect with your inner power. Power 12 is the flame that lights the energy of *Kukulkán*, and today his day is replete with the energies you need to increase your sexual vigor, develop your inner fire, and work on your evolution.

August
Thursday 9

13 Kame

Although we usually resist, change is good for us. You know which changes are the ones you need to make… Kame dissipates all things, and Power 13 guarantees success and grants you the power to alter your course. Today you can also get out of any bad business deal.

AUG

August
Friday 10

1 Kej

Enjoy the stability granted by the cycle of The Deer, as you will flourish no matter the direction in which you head! You will advance without encountering difficulties thanks to the *Nawal's* subliminal counsel and the solidarity provided by Power 1.

August
Saturday 11

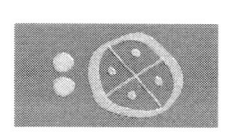

2 Q'anil

Today Power 2 symbolizes the couple and Q'anil contains the seeds of connubial love. This is a day of bliss and communion in which both partners can attain personal perfection, and gather the perfumed blossoms of intimacy, tenderness, and passion.

August
Sunday 12

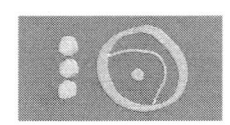

3 Toj

Power 3 opens the doors of communication with *B'itol* and *Tz'aqol* through the portal of the Sacred Fire, whose energies grant you the profound understanding that leads to the truth. Light a candle as a sign of repentance and payment for your transgressions.

August
Monday 13

4 Tz'i'

This is the ideal day in which to resolve complicated issues that need to be handled with transparency. You will be able to counteract any problem that involves the law, as Tz'i' represents accuracy and justice, and the truth will prevail with the support of Power 4.

AUG

August
Tuesday 14

5 B'atz'

Life is a resplendent tapestry of light and shadows! B'atz' places the threads at your disposal, but it is up to you to choose those that suit you best to weave the story of your life. Make sure your personal history displays triumphant and happy colors! Power 5 provides luck.

August
Wednesday 15

6 E

The Path is steep and uphill today, as Power 6 can weigh it down with trials and conflict. Even so, and no matter how difficult the journey, don't become discouraged or lose sight of your final destination, as tomorrow is another day and the view is glorious from the summit!

August
Thursday 16

7 Aj

On the day of Aj, the sacred maize was domesticated. Ask the *Nawal* to grant you the inner strength and endurance of the column, so that you can sustain the spiritual and material wellbeing of your home. Power 7 grants you faith in yourself.

August
Friday 17

8 I'x

The subtle duality of feminine intuition and masculine strength prevails on this day. The energy of Power 8 grants you vigor and authority when formulating personal or business strategies and I'x wields its sortilege to expand your mind.

AUG

August
Saturday 18

9 Tz'ikin

The shimmering wings of The Quetzal bring you wealth, luck, and love, but the wise and brilliant energy of Power 9 requires as a condition that you help your fellowmen. This is a good day to accomplish all those pending errands.

August
Sunday 19

10 Ajmaq

Ancestral wisdom is at your disposal through Ajmaq. Use its energy to penetrate deeply into the space in your intellect that holds the answers to every mystery, as the cosmic energies of Power 10 are at your unconditional service today.

August Monday 20

11 No'j

No'j embodies the power of thought; call upon its energy to help you attain the state of mental clarity that creates true happiness, which is inseparable from your consciousness and does not depend on external events. Power 11 puts your convictions to the test.

August Tuesday 21

12 Tijax

Power 12 projects family unity and Tijax restores and harmonizes your mental, spiritual, physical, and emotional bodies to correct any problems you may have with other people in any of these planes, especially with your family.

AUG

August
Wednesday 22

13 Kawoq

13 Kawoq vibrates with the energy of abundance, realization, and strength in union. This trecena has brought you lessons and knowledge; express your gratitude by sharing what you've learnt, so that your good deeds ripple like waves of light across the Universe!

August
Thursday 23

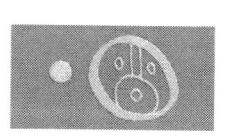

1 Ajpu

The Blowgun Hunter infuses you with the thrill of the hunt! Like all good hunters, you must practice patience, caution, and discipline to achieve your intentions during this cycle. Make sure that your goals are achievable. On its part, Power 1 offers innovative solutions.

August
Friday 24

2 Imox

Unusual and eccentric energies predominate today. It is a magnificent day for creative endeavors; explore new forms of expression and let your imagination flow without barriers! To inspire you, Power 2 creates a link between the material and the spiritual.

August
Saturday 25

3 Iq'

The petition you wish to make the Creator is strengthened today with the power that Iq' has of elevating your prayers to the highest. Power 3, which symbolizes communication, connects your energy to the Supreme Energies. Light a white candle and petition with faith.

AUG

August
Sunday 26

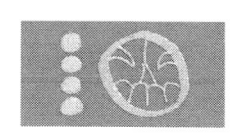

4 Aq'ab'al

Aq'ab'al, which is light and shadow, harbors the dark and luminous side we all carry within that influences the decisions we make every day. Use the energies of Power 4 so that harmony and balance prevail in everything you do today.

August
Monday 27

5 K'at

The energies of 5 K'at are working in your favor. On this day, you can catch all the dreams you have spun because, believe it, magic truly exists! Thank the Creator for the goals that you've accomplished so far and for everything you already have.

August
Tuesday 28

6 Kan

Watch your level of energy; the day will bring energetic highs and lows that could affect your vitality and decrease your physical and mental performance. Although Kan projects strength, Power 6 is up to its old tricks. Take the necessary precautions to avoid its effects!

August
Wednesday 29

7 Kame

Kame has the energy of our ancestors, who accompany us through life and when we return to our point of origin. Power 7 grants you the impulse you need to reach the top of the pyramid; use its energies for self-reflection.

AUG

August
Thursday 30

8 Kej

8 Kej is the day in which the Maya spiritual guides ask the Sacred Fire for guidance and the willingness and mental clarity to fulfill their calling. You too can ask the Sacred Fire for the spirit to serve others.

August
Friday 31

9 Q'anil

Q'anil harbors the dormant power of the seed and contains everything that is possible; sow love, health, bliss, material abundance, and spiritual wealth with its energy. Today you can rely on Power 9, the energy of realization, to activate these wonderful seeds!

September 2012

Sunday	Monday	Tuesday	Wednesday	Thursday	Friday	Saturday
						1 10 Toj
2 11 Tz'i'	3 12 B'atz'	4 13 E	5 1 Aj	6 2 I'x	7 3 Tz'ikin	8 ☽ 4 Ajmaq
9 5 No'j	10 6 Tijax	11 7 Kawoq	12 8 Ajpu	13 9 Imox	14 10 Iq'	15 11 Aq'ab'al
16 ● 12 K'at	17 13 Kan	18 1 Kame	19 2 Kej	20 3 Q'anil	21 4 Toj	22 ☾ **Fall Equinox** 5 Tz'i'
23 6 B'atz'	24 7 E	25 8 Aj	26 9 I'x	27 10 Tz'ikin	28 11 Ajmaq	29 12 No'j
30 ○ 13 Tijax						

September
Saturday 1

10 Toj

Power 10 connects you with Heart of the Sky and Heart of the Earth so that the Sacred Fire can cure the wounds from your past and you can find serenity. Ask the *Ajaw* to draw away the obstacles that keep you from developing your immanent powers.

September
Sunday 2

11 Tz'i'

Tz'i' emanates the energies of the Cosmic Authorities and Terrestrial Law, and together with the knowledge granted by Power 11, it reveals just how powerful your spiritual nature truly is. Energize yourself with the mystical power of the mountains!

SEP

September
Monday 3

12 B'atz'

B'atz' commands your past, present, and future, but things can't happen without your active intervention! Use the energy of Power 12 to materialize your wishes; if you can imagine it, you can create it.

September
Tuesday 4

13 E

You cannot transform yourself in one day, but today you reach a new level of perfection on your journey toward enlightenment. You are nearing the profound inner changes you desire; persevere, and you will attain it. Power 13 predicts success.

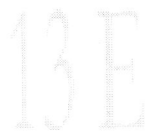

SEP

September
Wednesday 5

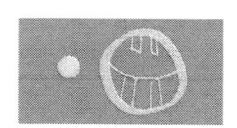

1 Aj

Dedicate this day to the matters that concern your home: your family, pets, and to all the living beings that give significance to and bring joy into your life. Devote this new cycle to them! Power 1 projects its unifying energy and Aj sustains the pillars of your home.

September
Thursday 6

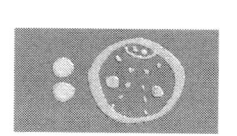

2 I'x

The Jaguar connects you to the power of the jungle and with the energies that develop superior powers. Your heightened perception will allow you to perceive the energies of others today, namely of that special person in whom you may have a romantic interest!

September
Friday 7

3 Tz'ikin

3 Tz'ikin sharpens your vision and intuition and rewards your efforts with results. Because of the foreknowledge it confers upon you, it is an excellent day to resolve any business issue or matter of the heart, as you will find the best solution to that particular situation.

September
Saturday 8

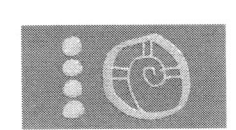

4 Ajmaq

Forgiveness is a spiritual act that requires valor and humility. This is a day for self-communion and to search profoundly into the purpose and direction of your life. Power 4 grants you the balance you need to overcome any hard feelings.

SEP

September
Sunday 9

5 No'j

No'j governs communication between dimensions and its energy will transmit to you knowledge and wisdom. It is a propitious day in which to carry out the ritual of divination: The sacred *tz'ite'* holds the answers, you just need to ask! Power 5 helps you carry out the changes you need to make.

September
Monday 10

6 Tijax

Today you can call upon the brilliant power of lightning to cut out any mysteries or illnesses, but be mindful of the way in which you manage the energy of this *Nawal*, as its conjunction with Power 6 could make you behave in a hurtful and stabbing manner toward those around you.

SEP

September
Tuesday 11

7 Kawoq

What can you do to help someone today? The Turtle radiates the energy of global consciousness and Power 7 contains the energy that encourages action. It is a privilege to give of yourself, and doing so will give you so much pleasure!

September
Wednesday 12

8 Ajpu

8 Ajpu infuses this day with a primordial, masculine, and enterprising vitality. Celebrate the father, brother, son, or friend; all the special men in your life! Today they will feel vibrant, and the day will flow with physical energy for them.

SEP

September
Thursday 13

9 Imox

Trust your intuition and let it guide you when making decisions on the important issues that will emerge today, as the energies of 9 Imox project the preeminence of the unconscious mind and wisdom. Don't disregard your inner voice!

September
Friday 14

10 Iq'

Open the windows of your mind so that the lucid ideas and positive energy of The Wind can enter! Its pure and clear force will guide you throughout this day with the support of Power 10, which will help you successfully carry out your projects.

September
Saturday 15

11 Aq'ab'al

Today Power 11 arrives with much learning for you, and it couldn't come at a better time, as Aq'ab'al grants you the opportunity to use these lessons to reinvent yourself! Use the knowledge you acquire to change all those things about you that you know you need to change.

September
Sunday 16

12 K'at

Use the energy of Power 12 to ignite a passion for what you do, and the tools provided by the *Nawal* to catch everything that you need to shine in your own right. You have so much talent, and your contribution is unique and special!

SEP

September
Monday 17

13 Kan

Ask the energies of 13 Kan to bring back the loved ones that have drifted away from your life. Everything that you thought was lost or forgotten will return to you, because the energies of the day create the ideal conditions.

September
Tuesday 18

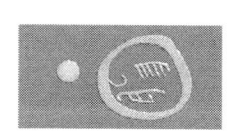

1 Kame

Kame dwells in the dimension of lasting peace and harmony. Light a candle to honor your ancestors, and ask them for their guidance and protection during this cycle. Power 1 and the *Nawal* offer you the opportunity to undertake profound inner changes.

SEP

September
Wednesday 19

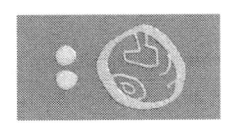

2 Kej

The stability that you seek can be found in the energy that flows from Kej, which balances your physical, mental, emotional, and spiritual manifestations and protects you from the feelings of anguish and chaos that life often entails; and in Power 2, which provides balance between positive and negative energies.

September
Thursday 20

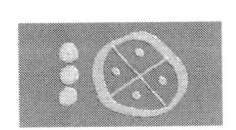

3 Q'anil

Q'anil contains the energy of creation and abundance within its sacred seed. Ask the *Nawal* to grant you knowledge in the use of medicinal and magical plants, as today Power 3 opens the portal that connects you with the energies of Universal Wisdom.

SEP

September
Friday 21

4 Toj

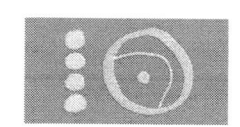

This is a special day in which to thank the Creator with an offering. Thank Him for all that you are and for everything that you have; for all the opportunities that you have received throughout your life, and for the life that still lies ahead. Light a white candle in direction to the east as a sign of gratitude.

September
Saturday 22

5 Tz'i'

Synchronize with the vibrations of the cosmos and Mother Earth! The Plumed Serpent shines a light between earth and heaven that awakens the radiance within you. The energy of the **equinox** is in harmony with the forces of Tz'i' and Power 5 and brings spiritual and material balance to your life.

September
Sunday 23

6 B'atz'

Connect with your creative and artistic side, as today The Monkey brings leisure and delight in everything that relates to the arts, although its conjunction with Power 6 could imply some trouble. Beware of its antics!

September
Monday 24

7 E

E provides the strength you will need to venture on the paths that will open before you along the course of your life. Invoke its energy so that the paths are level and bathed with the light of unforgettable experiences. Power 7 signals that you are nearing realization.

SEP

September
Tuesday 25

8 Aj

The cosmic and telluric pillars of Aj fill you with courage and vitality, and thanks to Power 8 the financial gains that you are awaiting could ensue. The *Nawal* grants you clairvoyance, and today you skillfully govern the elements of the material world with its gift.

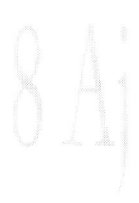

September
Wednesday 26

9 I'x

Feminine energy is at its peak! Today you are enveloped by the energies of love, spirituality, and the pure and perfect forces of creation. This is a day of high magic: The impossible is possible, and its energy grants you superior powers and intuition.

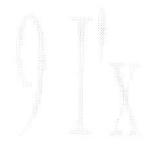

September
Thursday 27

10 Tz'ikin

Tz'ikin projects the energy of Heart of the Sky whose domain is the primordial silence that dwells throughout the Universe. Today you can free yourself of the constraints of time and space with the help of the cosmic energies so that global consciousness can flow within you.

September
Friday 28

11 Ajmaq

This is a day that is reserved for introspection. Look into your heart... What do you see? Today the energies of 11 Ajmaq will help cure the wounds that your misdeeds may have caused, and will allow you to reflect on the motives that prompt you to act in the wrong way.

SEP

September
Saturday 29

12 No'j

Wisdom and advice can reach you in many guises; learn to recognize when and how this happens. The *Nawal* grants you patience and prudence, and Power 12 provides you with the energy that will materialize your dreams. Aim for the stars!

September
Sunday 30

13 Tijax

You reach the end of this trecena and complete a cycle of profound learning. Use the power of Tijax to rid yourself of any negative energy you may have picked up along the way. Rub yourself down with one part salt, wash it off, then rub yourself with two parts sugar to do away with the unwanted vibrations.

SEP

October 2012

Sunday	Monday	Tuesday	Wednesday	Thursday	Friday	Saturday
	1 1 Kawoq	2 2 Ajpu	3 3 Imox	4 4 Iq'	5 5 Aq'ab'al	6 6 K'at
7 7 Kan	8 ☾ 8 Kame	9 9 Kej	10 10 Q'anil	11 11 Toj	12 12 Tz'i'	13 13 B'atz'
14 1 E	15 ● 2 Aj	16 3 I'x	17 4 Tz'ikin	18 5 Ajmaq	19 6 No'j	20 7 Tijax
21 8 Kawoq	22 ☾ 9 Ajpu	23 10 Imox	24 11 Iq'	25 12 Aq'ab'al	26 13 K'at	27 1 Kan
28 **Maximón feast** 2 Kame	29 ○ 3 Kej	30 4 Q'anil	31 5 Toj			

October
Monday 1

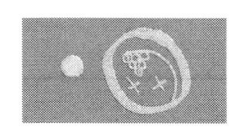

1 Kawoq

Don't wait around for others to do "something." 1 Kawoq displays strength in union and urges you to undertake actions that will benefit those around you. This trecena is calling you to action! Join those whose vision for a better world concurs with yours.

October
Tuesday 2

2 Ajpu

Ajpu takes aim and strikes the hearts of lovers everywhere! The energies of men and women vibrate in synchrony today, which sets the perfect conditions to nurture your relationship. Be understanding and loving to your partner; plan something fun and romantic!

OCT

October
Wednesday 3

3 Imox

Laugh, dance, sing, and above all, have fun! Today you are allowed to show your eccentric, outlandish, and whimsical self without worrying too much about appearances, so enjoy it fully but in moderation; a lack of control could lead you to experience strange events.

October
Thursday 4

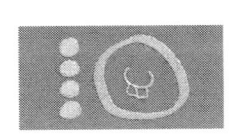

4 Iq'

Iq' unleashes a whirlwind of ideas that will make you spin 180 degrees in a completely unexpected direction! Luckily, Power 4 is here to protect you against any abrupt change and to rebalance you at a higher level.

OCT

October
Friday 5

5 Aq'ab'al

If you find that your efforts are leading you nowhere, today you can ask the *Nawal* to help you clarify your thoughts and to regain your bearings, as you can also rely on the energy of Power 5 to achieve anything you set your mind to accomplish.

October
Saturday 6

6 K'at

An energetic imbalance is possible given the confluence of Power 6 with The Net. Be alert at all times! You need to be very careful with everything you think, do or say today so as not to create or fall into any mental or circumstantial entanglements.

October
Sunday 7

7 Kan

Concentrate on the tone projected by the association of Power 7 with Kan, as their joint forces promise to provide you the initiative that you need. Plan your next steps so that you can move ahead from thought to action, and ultimately to completion.

October
Monday 8

8 Kame

Changes are foreseen with Kame whose energy is reinforced with the inherent strength and boldness that Power 8 has for transforming situations. Ask the Creator to put the people you need at this time on your path so that they can become your counselors and guides.

OCT

October
Tuesday 9

9 Kej

Seek contact with Mother Nature today through Kej so that she can nurture your mind, body, and spirit and lead you to that quiet place within. Power 9 will grant you the means to achieve this goal.

October
Wednesday 10

10 Q'anil

Just as your present is the result of your past actions use the cosmic energies of Power 10 to plant the seeds of wisdom and courage that you wish to harvest tomorrow. Stop and think about what it is that you really desire.

OCT

October
Thursday 11

11 Toj

You reach new evolutionary crossroads. The experience and learning gained so far allow you to recognize the consequences of your actions. You are granted the power to mend your conduct through The Sacred Fire, which can liberate you from the past.

October
Friday 12

12 Tz'i'

The strong and noble energy of Power 12 accompanies the *Nawal* throughout this day, which offers you the faculty to rectify any injustice that you may encounter. Light a candle and ask the Creator to help you always act with fairness.

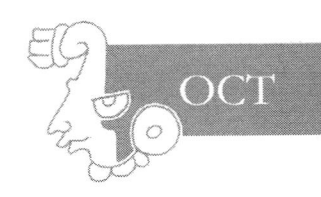

OCT

October
Saturday 13

13 B'atz'

The mystical connections you have woven together by the end of this cycle contain the energy of realization and the spiritual rewards that Power 13 confers upon you. Imbibe the wisdom they grant you, and wholeheartedly await the knowledge that is yet to come!

October
Sunday 14

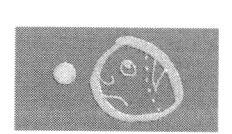

1 E

This is the cycle in which the portal opens broadly on the road ahead! Focus on what you wish to accomplish during this phase, as you can rely on the energies of the day to successfully carry out any of the projects you have in mind.

OCT

October
Monday 15

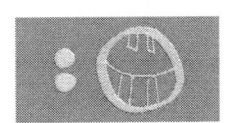

2 Aj

The Cane symbolizes the sacred energy that unites your family. Today it fuses with Power 2 so that you can lean on the strength and support of your loved ones when you need it. Be there for them as well.

October
Tuesday 16

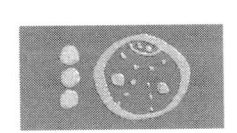

3 I'x

Channel the energy of The Jaguar to connect with your intuition, as today Power 3 will bind with the guiding forces that make your life flow without impediment. Pay attention to your inner voice!

OCT

October
Wednesday 17

4 Tz'ikin

Are you going through a period of emotional or financial unbalance? Today is the day in which you can change your situation, as Power 4 emanates balance and Tz'ikin fosters all aspects related with personal relationships and finances. Be confident; their energies are on your side.

October
Thursday 18

5 Ajmaq

Ajmaq confers the wisdom that allows you to explore and expand your consciousness. Concentrate on the energy of Power 5, which today reveals the love that flows from within you and lets you perceive the energy that radiates from all living creatures.

October
Friday 19

6 No'j

If you avail yourself to the patient and prudent energy of No'j you can overcome the inherent unbalance of Power 6. Be alert, as confusion may rule on this day. It is a good moment to quiet the mind and recharge your energies with a walk in the woods.

October
Saturday 20

7 Tijax

The Double Edged Knife consolidates its power today under the influence of Power 7. You could open advantageous opportunities, or undertake drastic or irreversible changes. These are powerful forces, so carefully ponder what you wish to accomplish with their energies.

October
Sunday 21

8 Kawoq

Channel the strength of this perfect day in which to attract the energy of abundance! Ask the Creator to grant you wealth, abundant love, and wisdom... or whatever you desire, since 8 Kawoq emanates the energies that will make your wishes come true!

October
Monday 22

9 Ajpu

The convergence of Power 9 with Ajpu projects an energy that is charged with vital and material forces. It is a day in which positive results are practically guaranteed. This is a good time to concern yourself with any pending projects you may have.

October
Tuesday 23

10 Imox

The telluric energies of Power 10 will keep you grounded in the face of the changing and eccentric powers of Imox. Likewise, energetic highs and lows can be expected on a day like today, in which you could also experience alternative realities.

October
Wednesday 24

11 Iq'

Can you hear it? The Wind is bringing you messages with the life lessons that you need for your development at this time. These lessons could be hard, so set up your mind to receive this knowledge with courage and acceptance.

October
Thursday 25

12 Aq'ab'al

Open up your dreamy eyes! Aq'ab'al peeks over the horizon and its brilliant rays shine on your most cherished fantasies, which come true with Power 12. Today you can distinguish the abstract from the concrete with amazing clarity.

October
Friday 26

13 K'at

Your wishes are like butterflies with resplendent wings, and today you have K'at to catch them! Without a doubt, you will develop the perfect plan thanks to the experience you've gained throughout the 13 Powers, so none will be out of your reach.

OCT

October
Saturday 27

1 Kan

Live passionately, as passion is the fuel that materializes all dreams. Awaken the fire of your youth and revive the aspirations you once had. Today the combined force of 1 Kan reestablishes your ability to take action.

October
Sunday 28

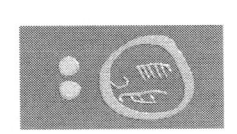

2 Kame

Today is the Feast of Maximón! Light eight yellow, white, red, and black candles and sprinkle white liquor over them so that the ethereal element can elevate your petitions. Maximón is powerful and understanding, and will help you resolve any problem you bring before him. On the day 2 Kame, you can put an end to any troublesome situations.

OCT

October
Monday 29

3 Kej

3 Kej guards you today from any lack of moderation on your part. You will find a perfect balance between work, fun, rest, and whatever impassions you! Your personal power will sustain you today on four pillars which are: truth, leadership, ethics, and justice.

October
Tuesday 30

4 Q'anil

Your imagination is such a fertile field that even your most bodacious ideas will flourish! Open the furrows of your mind to receive the solid, strong, and harmonious energies of 4 Q'anil so that they may all sprout, as everything that you start today will bring you joy.

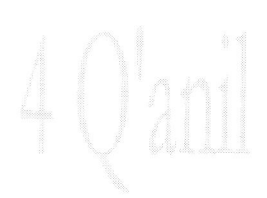

OCT

November 2012

Sunday	Monday	Tuesday	Wednesday	Thursday	Friday	Saturday
				1 **Day of the Dead** 6 Tz'i'	2 7 B'atz'	3 8 E
4 9 Aj	5 10 I'x	6 11 Tz'ikin	7 ☾ 12 Ajmaq	8 13 No'j	9 1 Tijax	10 2 Kawoq
11 3 Ajpu	12 4 Imox	13 ● 5 Iq'	14 6 Aq'ab'al	15 7 K'at	16 8 Kan	17 9 Kame
18 10 Kej	19 11 Q'anil	20 ☾ 12 Toj	21 13 Tz'i'	22 1 B'atz'	23 2 E	24 3 Aj
25 4 I'x	26 5 Tz'ikin	27 6 Ajmaq	28 ○ 7 No'j	29 8 Tijax	30 9 Kawoq	

October
Wednesday 31

5 Toj

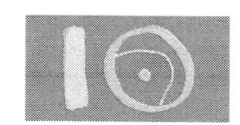

You are one of creation's most beautiful manifestations! The energy of The Sacred Fire governs the day; light a candle to connect with its divine energy and to harmonize with creation and thank the Creator for all that life has given you.

November
Thursday 1

6 Tz'i'

Power 6 presages discord and disruption. Don't go out the door without asking The Dog for his guidance and protection; he will faithfully assist you and lead you on the perfect path toward restoring harmony with the natural laws.

November
Friday 2

7 B'atz'

Destiny has brought you to this threshold. Gather strength for the final push to the summit, as B'atz' sheds new light on those unanswered questions rooted in your past. You may discover that you have known the answers all along.

November
Saturday 3

8 E

To get "there" you have to leave "here." Power 8 fuels you with uncommon courage and strength to travel de Sacred Road on the way finding your life's mission and fulfilling your destiny, as one step leads to the other.

NOV

November
Sunday 4

9 Aj

Your home is that still place in which you gather energy and regain your inner strength. Feel the power of Aj that sustains the pillars of your home, and the love, compassion, and feminine heart projected by Power 9, and rejoice!

November
Monday 5

10 I'x

Believe in magic! I'x possesses the power to expand your mind; but of course, it is up to you to open up to its energies. Change your point of view and then you will see… The cosmic and telluric energies of Power 10 will root you in reality but let your imagination soar.

November
Tuesday 6

11 Tz'ikin

Trials may be the order of the day thanks to Power 11, but on the bright side, Tz'ikin opens up new perspectives and frees you from the intellectual and materialistic straightjacket that you've put yourself into as of late.

November
Wednesday 7

12 Ajmaq

Ajmaq projects the energy of your aura; let it shine through, clear and dazzling! Power 12 grants you the faculty to acquire knowledge. Turn your eyes inward to see how far you have come on your road to spiritual evolution.

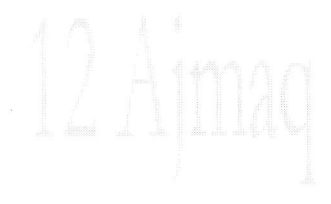

NOV

November
Thursday 8

13 No'j

Knowledge and experience that turn into wisdom converge today in the energies of 13 No'j. Transforming your consciousness is the most powerful thing you can do, for both yourself and humankind. How wonderful our world could be...!

November
Friday 9

1 Tijax

The Power of One will help you achieve significant changes in your life if you invoke the energies of Tijax to open up within you new depths of understanding. Together, these energies activate your comprehension of life's mysteries.

NOV

November
Saturday 10

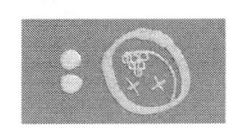

2 Kawoq

This day's energies project strength in union. Join hands and hearts with others in your community and seek to make it a wonderful and joyful place to live; you have so many talents and so much to offer!

November
Sunday 11

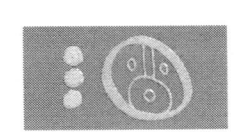

3 Ajpu

Call upon your inner spiritual warrior with the aid of the *Nawal* Ajpu, who can unburden you of the negative energies that keep you from advancing; and upon Power 3, which imbues you with the resourcefulness you need to sort things out.

NOV

November
Monday 12

4 Imox

The influence of Imox on the unconscious mind can lead to great confusion but, fortunately, the day is balanced with the Power of Four, which will help you fight off those unsteady forces and move you onward safely.

November
Tuesday 13

5 Iq'

On the one hand, Iq' the wind brings with it new ideas and the motivation you need to take action, and on the other, Power 5 provides freedom and signifies progress, so open up to all the wonderful possibilities!

November
Wednesday 14

6 Aq'ab'al

Today you might find it difficult to control the events around you; that sense of restlessness you are experiencing is perhaps Aq'ab'al's way of signaling that it's time to change course because you are entering a new cycle.

November
Thursday 15

7 K'at

Use the intuition granted by Power 7 to find a hole in the net that is tangling you up, and you can make an escape. Try to figure out how you got into that predicament in the first place, so you can learn from your mistakes.

November
Friday 16

8 Kan

The sexual enchantment of Kan and the physical power projected by Power 8 makes this a vital, magical day for men, who can be assured that their influence will be recognized and that excellent opportunities will arise in the material world.

November
Saturday 17

9 Kame

Life is uncertain, but on a day like today you can unequivocally rely on the guidance of your ancestors; life and death are part of the whole, and the connections between you cannot be severed. Power 9 transmits their loving energies to you.

November
Sunday 18

10 Kej

The cosmic energies are at your fingertips, and your feet are firmly planted on fertile soil thanks to the telluric energies emanating from deep within the heart of Mother Earth. Kej will see to it that nothing disturbs this balance today.

November
Monday 19

11 Q'anil

Experience is the teacher and transformation is the lesson. Make up your mind to let go of those old ideas and views, or your hands will be too full to receive the kernels of knowledge that this day has in store for you.

NOV

November
Tuesday 20

12 Toj

Toj is action and reaction; transgression and forgiveness. Be assured that your sincere contrition will be heard by the Father. Light a white candle to petition for His mercy and to materialize your dreams.

November
Wednesday 21

13 T'zi'

This trecena comes to an end under the guidance of the *Nawai* T'zi', who signals that authority and justice will be achieved on this day. Use the life experiences that you have gained throughout this cycle to counsel others.

November
Thursday 22

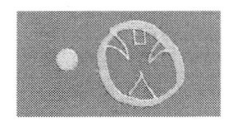

1 B'atz'

The high-powered confluence of Power 1 and B'atz' lend this day an unprecedented energy to embark on new beginnings, new projects, and to conceive new ideas. Everything about this day will be steeped in a shinning force that overflows with unconstrained originality and creativity!

November
Friday 23

2 E

This *Nawal* sets the conditions of your journey through life. Today it could lead you to a long-lasting and successful partnership, but ask for its guiding energy to overcome the polarity of Power 2.

November
Saturday 24

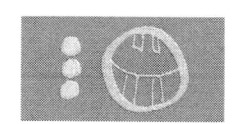

3 Aj

Your creativity and imagination will reach new heights thanks to Power 3. Today's energies bring forth excellence in everything you undertake, but Aj bids you to act with honesty and integrity in everything that you do and say.

November
Sunday 25

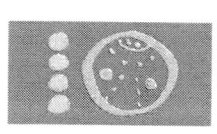

4 I'x

The wisdom of the four *Balam'eb* comes to you from the four corners of the Universe by means of Power 4, and I'x expands your level of consciousness to increase the energetic connection that exists between you and the Four Guardians.

NOV

November
Monday 26

5 Tz'ikin

The heaves are a vast open space; navigate them on the wings of Tz'ikin and open your heart to the idealism that inspires everything you do today. You will reach the highest peaks of any adventure you embark on today, as Power 5 grants you liberty and luck.

November
Tuesday 27

6 Ajmaq

Deepen your capacity to reflect with the rays of Ajmaq, and ask the *Nawal* to shed light on how to ask the *Ajaw* —and anyone you have offended— for forgiveness so as not to repeat your mistakes, as Power 6 also entails change and responsibility.

NOV

November
Wednesday 28

7 No'j

Power 7 has led you half way up the path; you might come across a few setbacks on your way to the top, but The Coyote, whose glyph —*Kab'awil*— represents the incremental degrees of spiritual perfection, will grant you wisdom and prudence.

November
Thursday 29

8 Tijax

The energies of the physical world flow invincibly today! The Obsidian Knife and Power 8 join forces and provide you with a powerful weapon against material blockages; brandish their powers to truncate any financial setbacks or illnesses that could be affecting you.

December 2012

Sunday	Monday	Tuesday	Wednesday	Thursday	Friday	Saturday
						1 10 Ajpu
2 11 Imox	3 12 Iq'	4 13 Aq'ab'al	5 1 K'at	6 ☾ 2 Kan	7 3 Kame	8 4 Kej
9 5 Q'anil	10 6 Toj	11 7 Tz'i'	12 **Waqxaqi' B'atz'** 8 B'atz'	13 ● 9 E	14 10 Aj	15 11 I'x
16 12 Tz'ikin	17 13 Ajmaq	18 1 No'j	19 2 Tijax	20 ☾ 3 Kawoq	21 **Winter Solstice** 4 Ajpu	22 5 Imox
23 6 Iq'	24 7 Aq'ab'al	25 8 K'at	26 9 Kan	27 10 Kame	28 ○ 11 Kej	29 12 Q'anil
30 13 Toj	31 1 Tz'i'					

November
Friday 30

9 Kawoq

What a blissful day is 9 Kawoq! Elevate your prayers to the Creator so that He may pour protection and blessings on your home and loved ones, as today the energies of the *Nawal* and the Tone envelop you in their love and contentment.

December
Saturday 1

10 Ajpu

Go ahead! Take that decision without any fear of making a mistake, as the energies of the day grant you certainty. Ajpu provides you with the cunning to overcome any adverse forces, and your day will flow with the powerful impulse of Power 10.

DEC

December
Sunday 2

11 Imox

It is possible that pessimistic thoughts will cross your mind today. Try to stay alert to this situation, as these ideas tend to manifest in the material world and can cause physical wear and tear. It won't be easy, but try to visualize things in a positive light.

December
Monday 3

12 Iq'

A whirlwind of beautiful, kind, and harmonious energies keep you company today; feed your spirit with this gift! Power 12, which represents the consciousness of being, offers you the opportunity to see yourself reflected in every person you encounter.

DEC

December
Tuesday 4

13 Aq'ab'al

Aq'ab'al brings to light all that is hidden from your intellect. Thanks to the insight your have gained through Power 13, you are prepared to face anything that this new cycle can bring. Seek the truth.

December
Wednesday 5

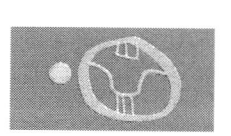

1 K'at

Throughout the course of this trecena, K'at will give you the tools that you need to consolidate those ideas that have come to mind and the people that can help you develop them. Power 1 grants you vigor and creative solutions.

December
Thursday 6

2 Kan

This day rushes in with the passion and impetuosity of a volcano. The life force that flows from *Kukulkán* joins Power 2 and brings you a day that is filled with a kind of vitality that is best enjoyed with your partner. Surrender to its energies!

December
Friday 7

3 Kame

Today you have a special connection with your ancestors, since Kame opens the channels between you to reveal their ancestral wisdom. Power 3, which symbolizes communication, unites the Three Cosmic Planes with your consciousness.

DEC

December
Saturday 8

4 Kej

The energy of stability reigns supreme today! Any project that you start will enjoy the protection of Kej, which provides the sturdiness of its four legs, and the balance of Power 4 and its powerful influence over the physical, mental, material, and spiritual manifestations.

December
Sunday 9

5 Q'anil

Ask the Creator to bring back into your life all that you once thought was lost, be it an old love, your health, or a project; or to resolve any unfinished situation, as The Seed blooms with possibilities and Power 5 provides realization.

DEC

December
Monday 10

6 Toj

This day is governed by Toj, which is the day in which to make offerings and attain reconciliation, and by Power 6, whose energy will provide lessons. The Sacred Fire lets you communicate with the Creator; ask Him to forgive you and to grant you light, peace, and discernment.

December
Tuesday 11

7 Tz'i'

The Count of the *Nawals* reaches anew its 260 day run. Use the spiritual force of Power 7, and the spiritual and material authority of Tz'i', to thank the Creator for the lessons, joy, and rewards that you have received throughout this period.

DEC

December
Wednesday 12

8 B'atz'

Maya New Year Celebration. Today is the first day of the *Cholq'ij* Sacred Calendar. During this day the *Ajq'ijab* —spiritual guides— and the members of the Maya community, meet to perform a Ceremony of Gratitude to the *Ajaw* and to ask Him for blessings for this new cycle. Light 20 candles and connect through their energy with the hundreds of Sacred Fires that are lit today in all of the Sacred Maya Altars.

December
Thursday 13

9 E

On the road to personal realization you will reach many junctions throughout your life, but Power 9 grants you the intuition to know which direction to take, and the strength to stay the course. This day encompasses a new and unexplored destination; live it to the fullest!

DEC

December
Friday 14

10 Aj

Firmly plant your feet on Mother Earth and elevate your arms so that your hands can grasp the cosmos. Feel the energies coursing through your body! The Pillar connects the cosmic with the terrestrial, and today you can connect with your center through Heart of the Sky and Heart of the Earth.

December
Saturday 15

11 I'x

You will be able to control the emotional highs and lows projected today by Power 2 if you focus on the power that I'x has to transform adverse circumstances. It is a good day to distance yourself from your daily activities to draw up a new life strategy.

DEC

December
Sunday 16

12 Tz'ikin

Thank the Creator for all the material assets you possess, as well as the money that reaches your hands. Ask Him for protection for your business, professional accomplishment, and material abundance. Light a pale blue candle to ask for opportunities, and a green one so that they are realized.

December
Monday 17

13 Ajmaq

13 Ajmaq indicates that you have completed a journey that gives you a profound sense of satisfaction, as the lucid state of your mind allows you to appreciate the degree of spiritual perfection that you have managed to achieve. Give yourself a pat on the back!

DEC

December
Tuesday 18

1 No'j

The mind, ideas, and wisdom govern the cycle presided by No'j, whose clarity you can use to harmonize the personal or labor relationships that you encounter during this trecena. This is a good day to resolve any problems with your partner.

December
Wednesday 19

2 Tijax

Use the force of Tijax to cut out of your life any bad luck, sickness, or enemies, as Power 2 creates a counterbalance between the positive and the negative that helps you to find the balance you need to attain physical and emotional wellbeing.

DEC

December
Thursday 20

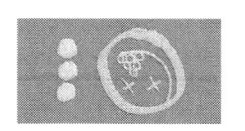

3 Kawoq

Appeal to the power that 3 Kawoq has to unite families, communities, and countries under its inclusive and optimistic energy, so that nations can find a road to peace and abundance for all people on earth. Let us invoke fraternal love!

December
Friday 21

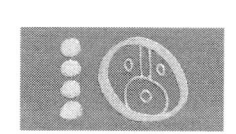

4 Ajpu

This is the day we have awaited! A 5200-year cycle in the *Ab'* calendar comes to an end as we enter an era that marks the beginning of a profound transformation in our collective consciousness and the end of materialism. You, spiritual warrior, charge yourself with the energies of the **solstice** to activate you inner powers!

December
Saturday 22

5 Imox

Put your mind in a receptive mode. Imox exercises its delicate effect over the inner spaces of your mind so that your creativity and intrinsic powers can flourish. Today the Tone strengthens spiritual reflection, and you can deftly manage the energies of change.

December
Sunday 23

6 Iq'

Windstorms are whipping up! Negative thoughts and emotional havoc may arise throughout the day from the unstable energies of 6 Iq', although with patience and reflection they could leave you a good dose of learning.

DEC

December
Monday 24

7 Aq'ab'al

Ask Aq'ab'al to dissipate the darkness with its light and to clarify which course will bring you new opportunities, as you are nearing the crest and you will need all the help it can give you to elucidate your next steps.

December
Tuesday 25

8 K'at

Today you could become entangled in earthly matters, and will have to make an effort to find balance between material and spiritual concerns. Ask K'at to carefully guard your memories, so that you can recapture them during the meditative moments of your life.

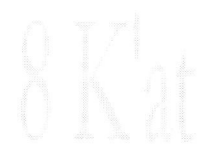

DEC

December
Wednesday 26

9 Kan

Nothing is stopping you today from fulfilling your dreams. Kan —the Creator and Maker of the Universe— instills its physical energy and wisdom in you, and Power 9 will additionally sharpen your intuition and grant you initiative. No more excuses!

December
Thursday 27

10 Kame

Power 10 opens a channel today that connects you with the Supreme Being, and Kame —the energy of life, death, and reincarnation— mediates between this world and the afterworld. Light incense and candles to honor the memory of your loved ones that have parted to the other dimension.

DEC

December
Friday 28

11 Kej

The wind, fire, earth, and water are sending you messages today through Kej, and they urge you to keep vigil on our planet's life-giving resources. The lessons you will experience today through Power 11 could closely relate to the four elements.

December
Saturday 29

12 Q'anil

This is a good day to ask the prevailing energies to grant you the capacity to overcome any lack of confidence you may have in yourself. 12 Q'anil harbors the energy of creation and your place within it. Your voice also deserves to be heard; express yourself fearlessly!

December
Sunday 30

13 Toj

Grandfather Sun is represented on earth through the *Tojil*, the Sacred Fire. Offer up some candles to the west, and thank Him for this new opportunity to feel, say, do, and undo. Power 13 shines a light on your gifts. Share them!

December
Monday 31

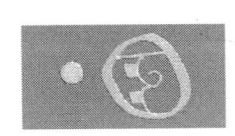

1 Tz'i'

Nothing can remain hidden for long from Tz'i', and justice that brings truth to light will prevail in this cycle. Ask the *Nawal* to exercise its power over terrestrial authorities everywhere, so that they can accomplish their duties with mysticism and vision. Power 1 fosters significant changes.

DEC

2012 Calendar of Auspicious and Inauspicious Days

AUSPICIOUS days for Men

Jan 9 / Sep 25	8 Aj	A day to acquire courage; it develops fortitude and reaffirms masculine authority
Feb 17 / Nov 3	8 E	Opportunities for negotiation, commerce, and business trips open up on this day
Apr 23	9 Tijax	A day to develop spiritual sensitivity and cut negative energies
Jul 9	8 Tz'ikin	An auspicious day for improving the financial situation of men, and developing, planning, and presenting projects

INAUSPICIOUS days for Men

Jan 10 / Sep 26	9 I'x	This day brings energetic highs and lows and confrontation between spouses
Feb 15 / Nov 1	6 Tz'i'	Can bring problems and confrontation with authorities
Feb 16 / Nov 2	7 B'atz'	Personal relationships or partnerships may become difficult on this day
Aug 31	9 Q'anil	Illnesses may arise today; children may feel misunderstood
Sep 2	11 Tz'i'	Possible legal complications or injustice may occur on this day

AUSPICIOUS days for Women

Jan 10 / Sep 26	9 I'x	A good day to develop artistic talents, magic, and to explore spiritual matters
Mar 18 / Dec 3	12 Iq'	Brings intellectual growth; an excellent day for project planning
Mar 19 / Dec 4	13 Aq'ab'al	A day in which to outline any changes women may want in their lives
Aug 17	8 I'x	This day brings material development and strength for women

INAUSPICIOUS days for Women

Mar 13 / Dec 28	7 No'j	This day can bring emotional stress to women
Mar 30 / Dec 15	11 I'x	Causes unbalance that especially affects women's emotional wellbeing
May 17	7 Iq'	Predicts emotional disruption and predisposes women to negative, unfounded thoughts
May 29	6 I'x	Many types of confrontations can surge today
Jul 8	7 I'x	Predicts emotional disruption and predisposes women to negative, unfounded thoughts
Jul 22	8 Q'anil	Illnesses may arise today, but it is a good day to ask for their withdrawal

AUSPICIOUS days for Children and Youths

Jan 13 / Sep 29	12 No'j	An intellectually stimulating day for children; teaches them nobility
Apr 22	8 No'j	Auspicious day for gaining knowledge at all levels; stimulates intellectual growth
May 10	13 Tz'ikin	Provides happiness and delight, especially in children younger than 7 years
Jul 10	9 Ajmaq	This day creates magic and stimulates curiosity that leads to inquiry and learning
Jul 31	4 No'j	Auspicious day for gaining knowledge at all levels; stimulates intellectual growth

INAUSPICIOUS days for Children and Youths

Jan 7 / Sep 23	6 B'atz'	This energy can generate bad relationships with teachers and classmates
Feb 3 / Oct 20	7 Tijax	An energy that brings insecurity; children may feel they need protection
Jul 25	11 B'atz'	During this day children may be prone to illness
Sep 11	7 Kawoq	Can cause conflicts with the family

AUSPICIOUS days for Seniors

Mar 6 / Nov 21	13 Tz'i'	A good day for seniors to petition for health and acquire energy
Mar 27 / Dec 12	8 B'atz'	A good day to petition for the recovery of the health of grandparents and seniors
May 22	12 Kej	A day in which elders can harmonize their lives with the lives of others
Jul 1	13 Kej	This day brings peace and tranquility to older adults

INAUSPICIOUS days for Seniors

Mar 17 / Dec 2	11 Imox	This day may have a negative effect on the emotional state, mind, and health of seniors
May 5	8 Tz'i'	A day that energetically has a detrimental effect on women
Jun 14	9 Tz'i'	A day that energetically has a detrimental effect on men

AUSPICIOUS days for Marriage

Jan 27 / Oct 13	13 B'atz'
Feb 24 / Nov 10	2 Kawoq
Mar 7 / Nov 22	1 B'atz'
Apr 16	2 B'atz'
May 6	9 B'atz'
Jul 5	4 B'atz'
Dec 12	8 B'atz'

INAUSPICIOUS days for Marriage

Jan 7 / Sep 23	6 B'atz'
Jan 25 / Oct 11	11 Toj
Mar 25 / Dec 10	6 Toj
Jul 25	11 B'atz'

All Kame and Tijax days

AUSPICIOUS days for the Home, Family Harmony, and Family Finances

Jan 31 / Oct 17	4 Tz'ikin	A day that brings stability in all aspects
Mar 11 / Nov 26	5 Tz'ikin	A good day to execute family plans, and to solve family problems or problems with the spouse
Jun 19	1 Tz'ikin	A day in which to revive love among the family, and reorganize domestic finances
Jun 23	5 Kawoq	Improve family finances and solve family problems with this energy
Jul 9	8 Tz'ikin	An energy that strengthens love and attracts financial luck
Aug 22	13 Kawoq	This day radiates beneficial energies for the family and abundance
Sep 5	1 Aj	All types of problems can be solved today if composure is kept

INAUSPICIOUS days for the Home, Family Harmony, and Family Finances

May 8	11 Aj	This day can bring conflict into the home
Aug 2	6 Kawoq	A conflictive energy; confrontations should be avoided so as to not increase any problems
Aug 16	7 Aj	An energy that can bring fear and doubt into the home and between partners
Sep 11	7 Kawoq	A likely day for confrontation and unbalance

AUSPICIOUS days to Renovate, Build, or Buy a Home / to Move

Jan 9 / Sep 25	8 Aj	This energy brings solidity and can help you find a good home
Feb 18 / Nov 4	9 Aj	Auspicious energies to find a new home that will generate much love and harmony
May 2	5 Kej	A good day to bring about harmony, good taste, and art into your home, to make it more comfortable and appealing
May 18	8 Aq'ab'al	A good day to remodel or change homes, as all changes will be lasting
Aug 10	1 Kej	An excellent day to find the home of your dreams

INAUSPICIOUS days to Renovate, Build, or Buy a Home / to Move

Jan 25 / Oct 11	**11 Toj**	Brings difficulties related with the construction or purchase of homes, or if moving
Mar 25 / Dec 10	**6 Toj**	This is not a good day to move into a new home!
May 4	**7 Toj**	A day that could bring a lack of consensus regarding the purchase or construction of a home, or about moving
Sep 2	**11 Tz'i'**	Legal conflicts could arise in relation with these issues

AUSPICIOUS days to ask for Physical, Mental or Spiritual Healing / to drive away Negativity and Depression / to undergo Cures

Jan 14 / Sep 30	**13 Tijax**	A good day to close a negative cycle
Mar 14 / Nov 29	**8 Tijax**	An excellent day for men
Apr 23	**9 Tijax**	An excellent day for women
Apr 27	**13 Iq'**	This day is ideal for calming the mind
Jun 6	**1 Iq'**	A good day to put an end to negative thoughts
Jun 18	**13 I'x**	A good day to drive away bad magic
Aug 5	**9 Iq'**	Beneficial energies abound today to drive away depression

There are no inauspicious days in this category

AUSPICIOUS days to ask for Guidance and Purpose in Life

Jan 31 / Oct 17	**4 Tz'ikin**	A good day in which to shore up any personal weaknesses
Mar 11 / Nov 26	**5 Tz'ikin**	A good day to seize courage and outline projects

Mar 28 / Dec 13	9 E	This day brings magic that can help you discover your path
May 18	8 Aq'ab'al	Ask for the surge of new opportunities on this day
Jul 9	8 Tz'ikin	A good day for material realization, and to petition for your finances
Aug 18	9 Tz'ikin	A day to communicate with the Supreme Beings and ask for guidance

There are no inauspicious days in this category

AUSPICIOUS days for Opening New Pathways / Finding Solutions

Jan 28 / Oct 14	1 E	E days can open positive spaces, especially these mentioned
Feb 17 / Nov 3	8 E	
Mar 28 / Dec 13	9 E	

INAUSPICIOUS days for Opening New Pathways / Finding Solutions

Jan 20 / Oct 6	6 K'at	This day brings conflict and confrontation
Jul 12	11 Tijax	A day with energies that sever the possibility of reaching the desired goals
Aug 7	11 K'at	This day's energies generate unbalance and confusion
Sep 10	6 Tijax	Conflicts that can lead to the loss of relationships or beneficial situations are possible today

AUSPICIOUS days for Developing Inner Powers / for Good Luck and Prosperity

Jan 10 / Sep 26	9 I'x	Women can find their connection between inner and outer energy
May 10	13 Tz'ikin	The noble energy of this day affords a connection with Superior Energies and the development of inner powers, especially vision and intuition

Jul 9	8 Tz'ikin	Men can find their connection between inner and outer energy
Aug 4	8 Imox	Men may develop vision and intuition, and a connection to their inner powers
Aug 17	8 I'x	Men can find their connection between inner and outer energies
Aug 18	9 Tz'ikin	Women can find their connection between inner and outer energies
Sep 13	9 Imox	Women may develop vision and intuition, and a connection to their inner powers

AUSPICIOUS days to drive away Bad Vibrations / perform Ritual Cleansings for the Home or Business / Personal Ritual Cleansings / drive away Harmful Persons / Revert Spells

Jan 14 / Aug 30	13 Tijax	A good day to work on any of the aspects mentioned in this category
Jan 22 / Oct 8	8 Kame	This day brings all cycles to a close
Mar 2 / Nov 17	9 Kame	Drive away strong energies and spells on this day; used only in complicated cases
Mar 14 / Nov 29	8 Tijax	A day that works well to drive away and slash material and financial blockages
Apr 23	9 Tijax	An especially good day to drive away hate, envy, revenge, or spells
Jun 22	4 Tijax	A day in which to drive away bad energies that emanate from people or places

There are no inauspicious days in this category

AUSPICIOUS days for Introspection, Meditation, and Spiritual Retirement

Apr 1 / Dec 17	13 Ajmaq	A day in which you may find the answers to your questions
Apr 22	8 No'j	The energies of this day help achieve mental balance
May 17	7 Iq'	This day grants energies that can help you attain a realistic attitude
Aug 5	9 Iq'	A good day to develop mental fortitude

INAUSPICIOUS days for Introspection, Meditation, and Spiritual Retirement

Jan 20 / Oct 6	6 K'at	This is a confusing day; it can generate confrontation and entanglements
Mar 17 / Dec 2	11 Imox	Instead of clarity, contradictory ideas may surge on this day
May 16	6 Imox	Unbalanced ideas and confrontation may surge today
Jun 11	6 Kej	This day may make you feel victimized
Aug 7	11 K'at	The unbalanced energies on this day may lead you to reach wrong conclusions

AUSPICIOUS days to Plan and Organize your Personal Finances

Jan 9 / Sep 25	8 Aj	Clarity and organizational abilities are the distinctive qualities of this day
Feb 17 / Nov 3	8 E	Open pathways and materialize good financial outcomes with this energy
Mar 11 / Nov 26	5 Tz'ikin	Grants the ability to explore and analyze business opportunities; attracts good fortune
Aug 22	13 Kawoq	A good day to plan and organize business aspects with partners; this day brings abundance

INAUSPICIOUS days to Plan and Organize your Personal Finances

Jan 17 / Oct 3	3 Imox	This day can generate unreasonable ideas that produce instability
Apr 12 / Dec 28	11 Kej	Treason may occur under the influence of this energy
Apr 21	7 Ajmaq	An energy that creates unbalance; the right path is difficult to discern

AUSPICIOUS days for Business / to Plan and Start New Businesses / to Establish New Partnerships

Feb 17 / Nov 3	8 E	An energy that opens up material pathways
Mar 27 / Dec 12	8 B'atz'	A good day to generate money and to plan new businesses
Apr 16	2 B'atz'	A good day for existing partnerships and to establish new ones
May 18	8 Aq'ab'al	This day opens new opportunities in the business world
Jun 19	1 Tz'ikin	A very good day for planning; this energy grants intuition to help recognize opportunities
Jul 9	8 Tz'ikin	A day that foretells good fortune

INAUSPICIOUS days for Business / to Plan and Start New Businesses / to Establish New Partnerships

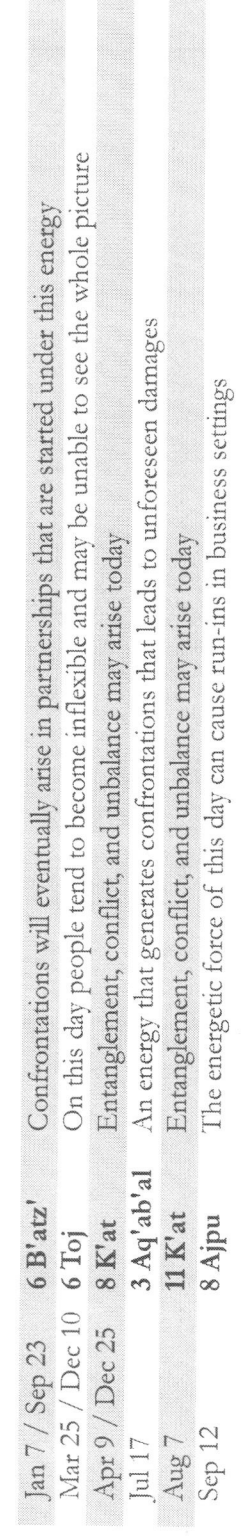

Jan 7 / Sep 23	6 B'atz'	Confrontations will eventually arise in partnerships that are started under this energy
Mar 25 / Dec 10	6 Toj	On this day people tend to become inflexible and may be unable to see the whole picture
Apr 9 / Dec 25	8 K'at	Entanglement, conflict, and unbalance may arise today
Jul 17	3 Aq'ab'al	An energy that generates confrontations that leads to unforeseen damages
Aug 7	11 K'at	Entanglement, conflict, and unbalance may arise today
Sep 12	8 Ajpu	The energetic force of this day can cause run-ins in business settings

AUSPICIOUS days to Conduct Legal Errands

Jan 26 / Oct 12	12 Tz'i'	A good day to reach a favorable divorce agreement
Mar 6 / Nov 21	13 Tz'i'	A good day in all respects for legal matters
May 5	8 Tz'i'	A good day to solve conflicts that concern financial and business matters

Jun 14	9 Tz'i'	A good day to solve conflicts that concern financial and business matters
Jun 28	10 K'at	A good day to help someone that may be imprisoned
Jul 13	12 Kawoq	A good day to reach a favorable divorce agreement

INAUSPICIOUS days to Conduct Legal Errands

Jan 20 / Oct 6	6 K'at	Things can become muddled and complicated today with this energy
Feb 15 / Nov 1	6 Tz'i'	The law is against you today and there can be confrontations
Apr 12 / Dec 28	11 Kej	Be aware: treason is forecasted on this day
Aug 7	11 K'at	Legal matters are unbalanced; it may be difficult to win a case under the influence of this energy
Sep 2	11 Tz'i'	A complicated day to achieve anything positive; the energy of the people that work in the legal profession is out of balance

AUSPICIOUS days to Ask for or Find a Partner / Start a New Romance

Jan 9 / Sep 25	8 Aj	Materialize the relationship with the help of this energy
Mar 11 / Nov 26	5 Tz'ikin	This favorable energy brings luck in love
Mar 27 / Dec 12	8 B'atz'	An energy that opens spaces for romantic encounters between couples
Apr 16	2 B'atz'	An energy that opens spaces for romantic encounters between couples
May 6	9 B'atz'	This energy grants attraction, love, and good feelings
Jun 19	1 Tz'ikin	This day opens spaces for romantic encounters

INAUSPICIOUS days to Ask for or Find a Partner / Start a New Romance

Jan 12 / Sep 28	11 Ajmaq	On this day you may meet someone very charming, but they may become quickly bored with the relationship, which inclines them toward infidelity
Mar 12 / Nov 27	6 Ajmaq	On this day you may meet someone very charming, but they may become quickly bored with the relationship, which inclines them toward infidelity
Apr 12 / Dec 28	11 Kej	This day brings unbalance and a possibility of treason
Jun 8	3 K'at	This energy brings stormy relationships
Jul 30	3 Ajmaq	On this day you may meet someone very charming but who may become quickly bored with the relationship, which inclines them toward infidelity
Aug 7	11 K'at	This energy brings stormy relationships

AUSPICIOUS days for Business or Pleasure Travel

Jan 31 / Oct 17	4 Tz'ikin	This day grants trips that bring tranquility and good relationships
Mar 31 / Dec 16	12 Tz'ikin	Pleasurable and exciting trips are forecasted on this day
May 27	4 E	Trips that begin on this day are restful and stable
Jun 19	1 Tz'ikin	Trips that begin on this day may bring interesting opportunities
Jul 9	8 Tz'ikin	Good business deals are possible, as the right person may cross paths with you today
Jul 10	9 Ajmaq	Adventures are in today's forecast
Jul 11	10 No'j	An ideal day to begin a rest and relaxation trip
Jul 26	12 E	This energy grants a restful and relaxed voyage
Aug 26	4 Aq'ab'al	Trips that begin on this day may bring interesting opportunities

INAUSPICIOUS days for Business or Pleasure Travel

Jan 20 / Oct 6	6 K'at	Mix-ups could arise today under the influence of this energy
Jan 22 / Oct 8	8 Kame	Traveling on this day can even be dangerous. The trip may be unexpectedly cancelled
Feb 20 / Nov 6	11 Tz'ikin	This day brings financial imbalance during the journey
May 21	11 Kame	An unbalanced day; trips begun on this day may bring negative changes
Jul 20	6 Kame	The trip may turn out to be filled with misfortune
Aug 7	11 K'at	Travel that starts on this day will not be fully enjoyed
Sep 10	6 Tijax	Trips made on this day harbor problems and annoyances

AUSPICIOUS days to Ask for and Find Work

Jan 9 / Sep 25	8 Aj	This day materializes the opportunity of a job
Jan 10 / Sep 26	9 I'x	With positive thinking, this energy creates the magic with which to attain any goal
Mar 19 / Dec 4	13 Aq'ab'al	An unlimited number of opportunities open up on this day
May 14	4 Kawoq	This energy fosters work in harmonious environments
Jul 27	13 Aj	The ideal conditions to improve your labor situation are created today
Sep 5	1 Aj	A number of opportunities are available to you today
Sep 12	8 Ajpu	This day brings certainty, and you may achieve what you desire

INAUSPICIOUS days to Ask for and Find Work

Jan 7 / Sep 23	6 B'atz'	You probably will not find acceptance from employers on this day
Jan 20 / Oct 6	6 K'at	Things become easily muddled due to today's energy
Feb 29 / Nov 15	7 K'at	Things become easily muddled due to today's energy

Mar 12 / Nov 27	6 Ajmaq	This energy complicates things today
May 8	11 Aj	This day may bring jobs that are unbalanced or cause adverse situations
Aug 7	11 K'at	Things become easily tangled-up with this energy

AUSPICIOUS days to Celebrate and Care for Mother Nature

Feb 26 / Nov 12	4 Imox	Imox days are beneficial for everything concerning the element water
Mar 7 / Nov 22	1 B'atz'	This is a good day to ask for all of Creation
Jun 19	1 Tz'ikin	A day to raise awareness of, and take action against, the destruction of nature
Jul 15	1 Imox	Develop your connection with Mother Earth on this day
Aug 10	1 Kej	A good day to plant, take care of, and raise awareness on plants, especially trees
Aug 30	8 Kej	An auspicious day to ask for all animals

Bibliography

Sam Colop, L.E. 2008. *Popol Wuj: Traducción al español y notas de Sam Colop*. Guatemala: Cholsamaj

Illustrations

The illustrations presented in this book are based on the ancient artistic writings of the Maya that are contained in the *Kumatzim Wuj Ka'i'*, also known as the Madrid Codex, and on those of the Codex Borbonicus.

Notes

Made in the USA
Middletown, DE
25 March 2017